Computational Techniques for Text Summarization based on Cognitive Intelligence

The book is concerned with contemporary methodologies used for automatic text summarization. It proposes interesting approaches to solve well-known problems on text summarization using computational intelligence (CI) techniques including cognitive approaches. A better understanding of the cognitive basis of the summarization task is still an open research issue; an extent of its use in text summarization is highlighted for further exploration. With the ever-growing text, people in research have little time to spare for extensive reading, where summarized information helps for a better understanding of the context in a shorter time.

This book helps students and researchers to automatically summarize text documents in an efficient and effective way. The computational approaches and the research techniques presented guides to achieve text summarization at ease. The summarized text generated supports readers to learn the context or the domain at a quicker pace. The book is presented with a reasonable amount of illustrations and examples convenient for the readers to understand and implement for their use. It is not to make readers understand what text summarization is, but for people to perform text summarization using various approaches. This also describes measures that can help to evaluate, determine, and explore the best possibilities for text summarization to analyze and use for any specific purpose. The illustration is based on social media and healthcare domain, which shows the possibilities to work with any domain for summarization. The new approach for text summarization based on cognitive intelligence is presented for further exploration in the field.

Computational Techniques for Text Summarization based on Cognitive Intelligence

V. Priya and K. Umamaheswari

CRC Press is an imprint of the
Taylor & Francis Group, an **informa** business

First edition published 2023
by CRC Press
6000 Broken Sound Parkway NW, Suite 300, Boca Raton, FL 33487-2742

and by CRC Press
4 Park Square, Milton Park, Abingdon, Oxon, OX14 4RN

CRC Press is an imprint of Taylor & Francis Group, LLC

ISBN: 9781032392820 (hbk)
ISBN: 9781032442471 (pbk)
ISBN: 9781003371199 (ebk)

DOI: 10.1201/9781003371199

Typeset in Times
by Deanta Global Publishing Services, Chennai, India

Contents

Preface

People have traditionally utilized written papers to convey important facts, viewpoints, and feelings. New technologies have caused an exponential rise in document output generated because of growing technology. In social networks, markets, production platforms, and websites, a tremendous volume of messages, product reviews, news pieces, and scientific documents are created and published every day. Although often verbose for the readers, this unstructured material can be quite helpful. The most pertinent material has been succinctly presented, and the reader is exposed to the key ideas thanks to the use of summaries. The new field of automatic text summarization was made possible by advances in text mining, machine learning, and natural language processing. These methods allow for the automatic production of summaries that typically contain either the most pertinent sentences or the most noticeable keywords from the document or collection. For visitors to become familiar with the content of interest rapidly, it is essential to extract a brief but informative description of a single document and/or a collection.

For instance, a synthesized overview of the most important news aspects may be provided by the summary of a group of news articles on the same subject. In contrast, the summary of social network data can help with the discovery of pertinent details about a particular event and the deduction of user and community interests and viewpoints. Several automatic summarizing techniques have been put forth in recent years that are broadly categorized into extractive summarization and abstractive summarization techniques.

This book offers a thorough examination of the state-of-the-art methods to describe text summarization. For both extractive summarizing tasks and abstractive summary tasks, the reader will discover in-depth treatment of several methodologies utilizing machine learning, natural language processing, and data mining techniques. Additionally, it is shown how summarizing methodologies can be used in a variety of applications, including healthcare and social media domain along with the possible research directions and future scope.

The book comprises seven chapters and is organized as follows. Chapter 1 'Concepts of Text Summarization' gives a basic but detailed text representation based on ideas or principles of text summarization. A detailed discussion of the ideas and practical examples are included for clear understanding. Some exercises related to text representation models are given to practitioners in the domain.

Chapter 2 'Large-Scale Summarization Using Machine Learning Approach' covers the representation of text summarization based on machine learning problems such as classification, clustering, deep learning, and others. It also examines the complexities and challenges encountered while using machine learning in the domain of text summarization.

Chapter 3 'Sentiment Analysis Approach to Text Summarization' addresses sentiment-based text summarization. Sentiment extraction and summarization provide an alternative notion for summarizing in order to capture users' sentiments or sensations in isolation as a summary. The consequences of modeling sentiment over text summarization have been explained, along with a full illustration of all stages. Some of the issues with sentiment-based text summarizing techniques have been explored.

Chapter 4 'Text Summarization Using Parallel Processing Approach' elaborates on text summarization using parallel processing techniques. In this chapter, MapReduce-based techniques are discussed with illustrations. The chapter includes a real-world example, focusing on problem formulation for large-scale summarization. It then covers the principal difficulties in large-scale text summarization. It ends with particular observations and unresolved issues regarding a comprehensive summary.

Chapter 5 'Optimization Approaches for Text Summarization' deals with optimization-based techniques like ant colony optimization (ACO), a type of nature heuristic algorithm, which is the most promising approach for tackling challenging optimization problems in text summarization. The authors specifically go into particle swarm optimization (PSO), a type of nature heuristic algorithm, which is the most promising approach for tackling challenging optimization problems for text summarization with the cognitive approaches. Indeed, a perception exists that the issue of text summaries has not been formally addressed using multi-objective optimization (MOO) tasks, despite the fact that numerous objectives must be accomplished. So authors also focus on framing text summarization for MOO problems and finally concluded with the specific challenges in framing text summarization as an optimization problem.

Chapter 6 'Performance Evaluation of Large-Scale Summarization Systems' discusses how to evaluate automatic text summarization systems. There are two widely used approaches: they are intrinsic and extrinsic methods. Extrinsic methods which include evaluation based on the application of text summaries in information retrieval and text similarity identification are elaborated in detail. The chapter also concludes with some of the research challenges in the evaluation of text summaries.

Chapter 7 'Applications and Future Directions' elaborates on the application of text summarization in two domains: social networks and health care. The chapter then goes into more detail about how to model the text summarization problem. The scope of summarization systems in various applications is then covered. The final section of the chapter discusses the use of sentiment analysis in text summarization. For current applications that have been addressed in detail one after another, there is a significant difference between traditional and large-scale summarization systems. The future direction of large-scale summarization was then discussed.

The various illustrations and code for certain concepts and an application using Python are added to the appendix for a clear understanding of the users.

About This Book

Textual information in the form of digital documents quickly accumulates to create huge amounts of data. The majority of these documents are unstructured: they are unrestricted text and have not been organized into traditional databases. Processing documents is therefore a perfunctory task, mostly due to a lack of standards. It has thus become extremely difficult to implement automatic text analysis tasks. Automatic text summarization (ATS), by condensing the text while maintaining relevant information, can help to process this ever-increasing, difficult-to-handle, mass of information.

This book examines the motivations and different algorithms for ATS. The authors present the recent state of the art before describing the main problems of ATS, as well as the difficulties and solutions provided by the community. The book provides the recent advances in ATS, as well as the current applications and trends. The approaches are statistical, linguistic, and symbolic. Several examples are also included in order to clarify the theoretical concepts.

1 Concepts of Text Summarization

1.1 INTRODUCTION

Text summarization is a process of grouping key points of the text from a large collection of text sources. The collection of sources can be reviews from the web, documents, blog posts, social media posts, news feeds, etc. Generating a concise and precise summary of voluminous texts is the key idea in summarization. This enables the summarization systems to focus on sections that convey useful information without loss of original meaning. These systems aim to transform lengthy documents into shortened versions. The purpose is to create a coherent and fluent summary which outlines the mandatory information in the document. Two major methods to generate summaries are manual summary generation and automatic summary generation. Manually generating a summary can be time consuming and tedious. Automatic text summarization systems became more prevalent with the growth of data posted/uploaded in the form of text in World Wide Web in this cyberspace. Automatic text summarization is a common problem in the field of machine learning and natural language processing (NLP). With such a big amount of data circulating in the digital space, there is a need to develop machine learning algorithms that can automatically shorten longer texts and deliver accurate summaries that can fluently pass the intended messages. The need for automatic text summarization could be useful before proceeding to the various techniques.

1.2 NEED FOR TEXT SUMMARIZATION

In current digital world, data is exploded in huge volumes, which are mostly non-structured textual data and it becomes un-readable most often. So this necessitates the extraction of important concepts to make it readable. There is a need to develop automatic text summarization tools to perform extraction and summarization. This allows people to get insights into the context easily. Currently quick access to enormous amounts of information is enjoyed by most of the people. However, most of this information is redundant and insignificant and may not convey the intended meaning. For example, if somebody is looking for specific information from an online news article, they may have to dig through its content. Also they have to spend a lot of time weeding out the unnecessary text before getting into the relevant content. Therefore, automatic text summarizers capable of extracting useful information and filtering inessential and insignificant data become necessary in this information era. Implementing summarization can enhance the readability of documents and reduce the time spent researching

DOI: 10.1201/9781003371199-1

information. Some day-today tasks where summarization finds its applications are as follows:

- Headlines (from around the world)
- Outlines (notes for students)
- Minutes (of a meeting)
- Previews (of movies)
- Reviews (of a book, compact disc (CD), movie)
- Biography (resumes)

1.3 APPROACHES TO TEXT SUMMARIZATION

The two major approaches used for automatic text summarization are extractive text summarization and abstractive text summarization as shown in Figure 1.1.

1.3.1 Extractive Summarization

The extractive text summarization technique involves extracting key phrases from the source document and combining them to make a summary. The extraction is made according to the defined features and significant terms as a key metric, without making any changes to the texts. Extraction involves concatenating extracts taken from the corpus into a summary. It has been observed that in the context of multi-document summarization of news articles, extraction may be inappropriate because it may produce summaries which are biased towards some sources. Extractive techniques involve preprocessing and models for extracting significant phrases or words and then generating a summary.

1.3.2 Abstractive Summarization

The abstraction technique entails paraphrasing and shortening parts of the source document. When abstraction is applied for text summarization in common machine learning problems, it can overcome the grammar inconsistencies of the

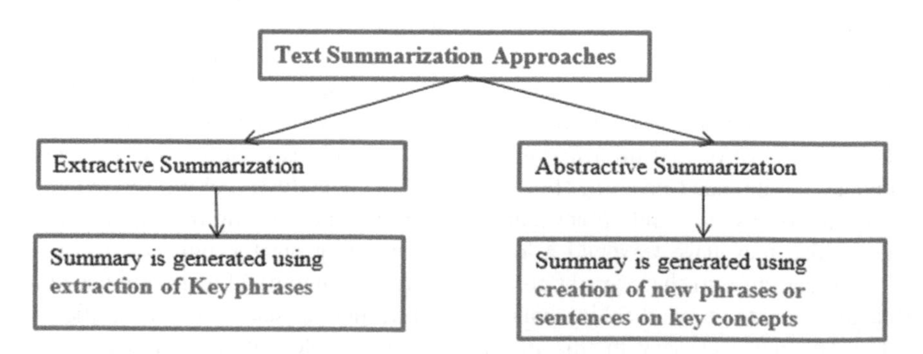

FIGURE 1.1 Text summarization approaches

extractive method. Abstraction involves generating novel sentences from information extracted from the corpus. The abstractive text summarization algorithms create new phrases and sentences that relay the most useful information from the original text – just like humans do. Abstraction involves complex techniques such as natural language processing, paraphrasing, and natural language generation methods.

Even though abstraction performs better than extraction, text summarization algorithms required to do abstraction use graphical representations and complex knowledge bases. This makes the use of extraction techniques still popular. In extraction methods, machine learning methods have proved to be very effective and successful in both single- and multi-document summarization.

The following section elaborates on the text modeling methods for extractive summarization.

1.4 TEXT MODELING FOR EXTRACTIVE SUMMARIZATION

Unstructured text requires precise representation or modeling, before being given to text summarization algorithms for a summary generation. Some of the text modeling methods like bag-of-words (BoW) model, vector space model (VSM), and text representation schemes have been detailed below.

1.4.1 Bag-of-Words Model

The bag-of-words model is a way of representing text data when modeling text for various real-world problems. The bag-of-words model is simple to understand and implement and has seen great success in problems such as natural language modeling and document classification. A problem with modeling text is that it is cluttered, and most of the algorithms prefer well-defined fixed-length inputs and outputs.

A bag-of-words model, or BoW for short, is a way of extracting features from text for use in modeling, such as with machine learning algorithms. The approach is very simple and flexible and can be used in a myriad of ways for extracting features from documents. A bag-of-words is a representation of text that describes the occurrence of words within a document. It involves two factors: a vocabulary of known words and a measure of the presence of known words.

It is called a 'bag' of words, because any information about the order or structure of words in the document is discarded. The model is only concerned with whether known words occur in the document and it discards about the context of the words present in the document.

Example for the bag-of-words model is given below:

Collection of Data

Below is a snippet of two lines of text from the reviews:

John likes to watch movies. Mary likes movies too.
Mary also likes to watch football games.

For this small example, treat each line as a separate 'document' and the two lines as the entire corpus of documents.

Designing the Vocabulary

Now we can make a list of all the words in our model vocabulary. The unique words here (ignoring case and punctuation) are:

'John'
'likes'
'to'
'watch'
'movies'
'too'
'also'
'football'
'games'
'Mary'

The above is the model vocabulary of ten words from a corpus containing 16 words.

Creation of Document Vectors

The next step is to score the words in each document and convert them into a vector. The document of free text into a vector is used as input or output for a machine learning model. As the vocabulary is built with ten words, it can use a fixed-length document representation of 10, with one position in the vector to score each word.

The simplest scoring method is to mark the presence or non-existence of words represented as a Boolean value, 0 for absent, 1 for present in the document. Using the arbitrary ordering of words listed in the above vocabulary, the first step to convert the document (John likes to watch movies. Mary likes movies too) into a binary vector is performed as follows:

The scoring of the document would look as follows:

'John' = 1
'likes' = 1
'to' = 1
'watch' = 1
'movies' = 1
'Mary' = 1
'too' = 1
'Also'=0
'football' =0
'games' =0
As a binary vector, this would look as follows: [1, 1, 1, 1, 1, 1, 1, 0, 0, and 0]

The next document would look as follows: 'Mary also likes to watch football games' = [0, 1, 1, 1, 0, 1, 0, 1, 1, and 1]

All ordering of the words is nominally discarded and has a consistent way of extracting features from any document in our corpus, ready for use in modeling.

Other scoring mechanisms that are in use are n-gram model, ti-idf, etc.

The bag-of-words model is very simple to understand and implement. This model offers a lot of flexibility for customization on your specific text data. It has been used with great success on prediction problems like language modeling and documentation classification.

Major limitations of BoW model are:

Semantic meaning: The basic BoW approach does not consider the meaning of the word in the document. It completely ignores the context in which it is being used. The same word can be used in multiple places based on the context or nearby words.

Vector size: For a large document, the vector size can be huge, resulting in a lot of computation, and also construction requires an enormous time.

Example for BoW model:

1.4.2 Vector Space Model

Vector space model was first proposed by Salton and Buckley for information retrieval. It uses vectors to represent documents, and the elements of a vector consist of words appearing in the collection. The mathematical representation is given in Equation 1.1 as follows:

$$v = \begin{bmatrix} v_{1,1} & v_{1,2} & v_{1,3} \\ v_{2,1} & v_{2,2} & v_{2,3} \\ . & . & v_{n,n} \end{bmatrix} \tag{1.1}$$

The rows of the matrix are defined as documents in the vector space while the columns of the matrix are defined as the terms which are used to describe or index the documents in the vector space. This matrix is commonly referred to as the document–term matrix. There are many term-weighting schemes used for representing words in a document. Commonly used among them and proven to be very effective is tf-idf scheme for term scoring. This has been first applied to relevance decision-making by Wu et al. (2008).

An example for vector space representation is shown below:

Consider three documents given below:

d1: 'new york times'

d2: 'new york post'

d3: 'los angeles times'

For the three documents, the idf score can be computed using the formula:
Idf = total no of documents / total number of times the term appears in the document
angles log2(3/1) = 1.584
los log2(3/1) = 1.584
new log2(3/2) = 0.584
post log2(3/1) = 1.584
times log2(3/2) = 0.584
york log2(3/2) = 0.584

The term frequency matrix is given by Table 1.1.

Multiplying tf by idf for all the terms, we obtain the tf-idf matrix as shown in Table 1.2.

Some of the limitations of tf-idf scheme are:

- The model computes document similarity directly in the word count space, which may be slow for large vocabularies.
- It assumes that the counts of different words provide independent evidence of similarity.
- It makes no use of semantic similarities between words.

Example 1.1

The bag-of-words (BoW) model is the simplest form of text representation in numbers. Like the term itself, we can represent a sentence as a bag-of-words vector (a string of numbers).

TABLE 1.1
Document Representation Using Vector Space Model

	Angeles	Los	New	Post	Times	York
D1	0	0	1	0	1	1
D2	0	0	1	1	0	1
D3	1	1	0	0	1	0

TABLE 1.2
TF-IDF Matrix for the Documents

	Angeles	Los	New	Post	Times	York
D1	0	0	0.584	0	0.584	0.584
D2	0	0	0.584	1.584	0	0.584
D3	1.584	1.584	0	0	0.584	0

Let's recall the three types of movie reviews we saw earlier:

Review 1: This movie is very scary and long.
Review 2: This movie is not scary and is slow.
Review 3: This movie is spooky and good.

We will first build a vocabulary from all the unique words in the above three reviews. The vocabulary consists of these 11 words: 'This,' 'movie,' 'is,' 'very,' 'scary,' 'and,' 'long,' 'not,' 'slow,' 'spooky,' 'good.'

We can now take each of these words and mark their occurrence in the three movie reviews above with 1s and 0s. Table 1.3 gives us three vectors for three reviews:

Vector of Review 1: [1 1 1 1 1 1 1 0 0 0 0]
Vector of Review 2: [1 1 2 0 0 1 1 0 1 0 0]
Vector of Review 3: [1 1 1 0 0 0 1 0 0 1 1]

We will again use the same vocabulary we had built in the bag-of-words model to show how to calculate the TF for Review 2:

Review 2: This movie is not scary and is slow.
Here,
Vocabulary: 'This,' 'movie,' 'is,' 'very,' 'scary,' 'and,' 'long,' 'not,' 'slow,' 'spooky,' 'good'
Number of words in Review 2 = 8
TF for the word 'this' = (number of times 'this' appears in review 2) / (number of terms in review 2) = 1/8
Similarly,
TF('movie') = 1/8
TF('is') = 2/8 = 1/4
TF('very') = 0/8 = 0
TF('scary') = 1/8
TF('and') = 1/8
TF('long') = 0/8 = 0

TABLE 1.3
Table Representing Three Vectors for Three Reviews

	1 This	2 Movie	3 Is	4 Very	5 Scary	6 And	7 Long	8 Not	9 Slow	10 Spooky	11 Good	Length of the Review (in Words)
Review 1	1	1	1	1	1	1	1	0	0	0	0	7
Review 2	1	1	2	0	0	1	1	0	1	0	0	8
Review 3	1	1	1	0	0	0	1	0	0	1	1	6

TF('not') = 1/8
TF('slow') = 1/8
TF('spooky') = 0/8 = 0
TF('good') = 0/8 = 0

We can calculate the term frequencies for all the terms and all the reviews as in Table 1.4.

We can calculate the IDF values for all the words in Review 2:

IDF('this') = log(number of documents / number of documents containing the word 'this') = log(3/3) = log(1) = 0
Similarly,
IDF('movie') = log(3/3) = 0
IDF('is') = log(3/3) = 0
IDF('not') = log(3/1) = log(3) = 0.48
IDF('scary') = log(3/2) = 0.18
IDF('and') = log(3/3) = 0
IDF('slow') = log(3/1) = 0.48

We can calculate the IDF values for each word as shown above.

Hence, we see that words like 'is,' 'this,' 'and,' etc. are reduced to 0 and have little importance while words like 'scary,' 'long,' 'good,' etc. are words with more importance and thus have a higher value.

We can now compute the TF-IDF score for each word in the corpus. Words with a higher score are more important, and those with a lower score are less important.

TF_IDF Formula

TABLE 1.4
Table Representing Calculation of Frequency for the Three Reviews

Term	Review 1	Review 2	Review 3	TF (Review 1)	TF (Review 2)	TF (Review 3)
This	1	1	1	1/7	1/8	1/6
Movie	1	1	1	1/7	1/8	1/6
Is	1	2	1	1/7	1/4	1/6
Very	1	0	0	1/7	0	0
Scary	1	1	0	1/7	1/8	0
And	1	1	1	1/7	1/8	1/6
Long	1	0	0	1/7	0	0
Not	0	1	0	0	1/8	0
Slow	0	1	0	0	1/8	0
Spooky	0	0	1	0	0	1/6
Good	0	0	1	0	0	1/6

We can now calculate the TF-IDF score for every word in Review 2:

TF-IDF('this', Review 2) = TF('this', Review 2) * IDF('this') = 1/8 * 0 = 0
TF-IDF('movie', Review 2) = 1/8 * 0 = 0
TF-IDF('is', Review 2) = 1/4 * 0 = 0
TF-IDF('not', Review 2) = 1/8 * 0.48 = 0.06
TF-IDF('scary', Review 2) = 1/8 * 0.18 = 0.023
TF-IDF('and', Review 2) = 1/8 * 0 = 0
TF-IDF('slow', Review 2) = 1/8 * 0.48 = 0.06

Similarly, we can calculate the TF-IDF scores for all the words with respect to all the reviews.

TF-IDF Scores

We have now obtained the TF-IDF scores for our vocabulary. TF-IDF also gives larger values for less frequent words and is high when both IDF and TF values are high, i.e. the word is rare in all the documents combined but frequent in a single document.

1.4.3 Topic Representation Schemes

Topic modeling is an unsupervised machine learning technique. It is capable of scanning a set of documents, detecting word and phrase patterns within them. Based on the patterns identified, it clusters word groups and similar expressions that best characterize a set of documents automatically. Topic modeling is an 'unsupervised' machine learning technique, in other words, one that does not require training. The technique automatically analyzes text data to determine cluster words for a set of documents. The method involves counting words and grouping similar word patterns to infer topics within unstructured data.

Topic modeling refers to the process of dividing a corpus of documents in two:

- A list of topics covered by the documents in the corpus
- Several sets of documents from the corpus grouped by the topics they cover

The underlying assumption is that every document comprises a statistical mixture of topics, i.e. a statistical distribution of topics that can be obtained by 'adding up' all the distributions for all the topics covered. The two major topic modeling methods, namely latent semantic analysis (LSA) and latent Dirichlet allocation (LDA), are detailed in the next section.

Latent Semantic Analysis

Latent semantic analysis (LSA) is one of the most frequent topic modeling methods used by many analysts. It is based on what is known as the distributional hypothesis which states that the semantics of words can be grasped by looking at the contexts the words appear in. In other words, under this hypothesis, the semantics of two words will be similar if they tend to occur in similar contexts.

All topic models are based on the same basic assumption: Each document consists of a mixture of topics, and each topic consists of a collection of words.

Latent semantic analysis computes how frequently words occur in the documents using the whole corpus. It assumes that similar documents will contain approximately the same distribution of word frequencies for certain words in the corpus. In this case, syntactic information (e.g. word order) and semantic information (e.g. the multiplicity of meanings of a given word) are ignored, and each document is treated as a bag of words.

The standard method for computing word frequencies is tf-idf. Once tf-idf frequencies have been computed, we can create a document–term matrix which shows the tf-idf value for each term in a given document. This matrix will have rows representing every document in the corpus and columns representing every term considered. This document–term matrix can be decomposed into the product of three matrices (U * S * V) by using singular value decomposition (SVD). The U matrix is known as the document–topic matrix, and the V matrix is known as the term–topic matrix: Linear algebra guarantees that the S matrix will be diagonal and LSA will consider each singular value, i.e. each of the numbers in the main diagonal of matrix S, as a potential topic found in the documents. Now, if the largest '*t*' singular values are kept together with the first '*t*' columns of U and the first *t* rows of V, then '*t*' frequent topics found in the original document–term matrix can be obtained.

The quality of the topic assignment for every document and the quality of the terms assigned to each topic can be assessed through different techniques by looking at the vectors that make up the U and V matrices, respectively.

Latent Dirichlet Allocation

LDA stands for latent Dirichlet allocation. LDA is a Bayesian version of probabilistic LSA. In particular, it uses Dirichlet priors for the document–topic and word–topic distributions, lending itself to better generalization.

Latent Dirichlet allocation (LDA) and LSA are based on the same underlying assumptions: the distributional hypothesis (i.e. similar topics make use of similar words) and the statistical mixture hypothesis (i.e. documents talk about several topics) for which a statistical distribution can be determined. The purpose of LDA is mapping each document in our corpus to a set of topics which covers a good deal of the words in the document.

Latent Dirichlet allocation maps the documents to a list of topics for assigning topics to arrangements of words, e.g. n-grams such as best player for a topic related to sports. This stems from the assumption that documents are written with arrangements of words, and those arrangements determines the key concepts of the listed topics. Just like LSA, LDA also ignores syntactic information and treats documents as bags of words. It also assumes that all words in the document can be assigned a probability of belonging to a topic. The goal of LDA is to determine the mixture of topics that a document contains. The main difference between

LSA and LDA is that LDA assumes that the distribution of topics in a document and the distribution of words in topics are Dirichlet distributions. LSA does not assume any distribution and therefore leads to more opaque vector representations of topics and documents.

1.4.4 Real-Valued Model

There are many word representations available which are modeled through real-valued feature vectors. They are capable of capturing global syntactic and semantic dependencies between words. The recent work on word embedding–based representation is described in this section.

Real-valued syntactic word vectors (RSV) is a method of word embedding that builds a set of word vectors from right singular vectors of a transformed co-occurrence matrix. RSV method has been modified by the same authors by using entropy-based feature selection and adaptive transformation. RSV method has been explained in detail below:

> RSV comprises three steps: context-word vectors, transformation, and dimensionality reduction. These are explained based on distributional semantic space. A distributional semantic space can be built in three main steps.
>
> First, each word is associated with a vector, called context-word vector. The elements of a context-word vector associated with a word are the frequencies of seeing the word in certain contexts. Context is a region in a text whose presence can be attributed to a set of textual units, called context units. The relationships between words and context units in certain contexts are measured by a function, called the local weighting function.
>
> In the second step, a transformation function is applied to the context-word vectors in order to weigh the importance of contexts in discriminating between words. We will refer to this function as a global weighting function.
>
> Finally, in the third step, a set of low-dimensional word vectors is extracted from these transformed context-word vectors through the application of dimensionality reduction techniques.

1.5 PREPROCESSING FOR EXTRACTIVE SUMMARIZATION

The importance of preprocessing procedure is evident because it is used in almost every developed system related to text processing. Preprocessing includes one or more of the following steps.

- The first step is lexical analysis of the text with the objective of treating digits, hyphens, punctuation marks, and the case of letters.
- In the second step, semantically empty words (stop words) can be eliminated.
- Finally, stemming can be applied which comprehends grouping similar words together by means of small changes: for example, eliminating

prefixes and suffixes of words. The task of extractive text summarization consists of communicating the most important information of a document.

Usually, linguistic, statistic, or heuristic methods are used to extract the most important parts of a text and use them as new terms. These terms must be more precise and shorter than original text. Moreover, the terms tend to reflect the content in the best possible manner. Such terms can be of different granularity: words or multiword terms such as n-grams or whole phrases. In this sense, the terms which are composed of several words can be called multiword descriptions.

Some of the common preprocessing tasks in text processing applications are:

- Lower casing
- Punctuation, stop words, frequent, and rare words removal
- Spelling correction

Tokenization

Tokenization is the process of dividing text into a set of meaningful pieces. These pieces are called tokens.

Lemmatization

Lemmatization is an approach to eliminate inflections in the natural language text. Inflections in a natural language refer to morphological variants of a verb. Thus, lemmatization process reduces the words with the same meaning to its correct base forms of words. The key to this methodology is linguistics. To extract the proper lemma, it is necessary to look at the morphological analysis of each word. This requires having dictionaries for every language to provide that kind of analysis. Lemmatization does not simply chop off inflections but instead relies on a lexical knowledge base like WordNet to obtain the correct base forms of words.

Original sentence:

The striped bats are hanging on their feet for best.

Lemmatized sentence:

The striped bat are hanging on their foot for best.

Stemming

Stemming algorithms work by cutting off the end or the beginning of the word, taking into account a list of common prefixes and suffixes that can be found in an inflected word. This indiscriminate cutting can be successful in some occasions, but there may be some limitations. There are different algorithms that can be used in the stemming process, but the most common in English is Porter stemmer or Stanford stemmer. Stemming algorithms are typically rule-based.

Two common issues in stemming are over-stemming and under-stemming. Over-stemming occurs when too much of a word is cut off. This results in a situation that the meaning of the entire word is lost or scrambled.

Examples: university, universal, universities, and universe.

A stemming algorithm that stems these four words to the stem 'univers'.

Under-stemming is the opposite issue. It occurs when two words are stemmed to the same root that are not of different stems.

All the words alumnus, alumni, and alumna are stemmed using the porter stemming algorithm as below:

'alumnus' → 'alumnu', 'alumni' → 'alumni', 'alumna'/'alumnae' → 'alumna'

Applications of stemming are:

- Stemming is used in information retrieval systems like search engines.
- It is used to determine domain vocabularies in domain analysis.

Part of Speech Tagging

Part of speech tagging or POS tagging is the common processing technique where the words in a sentence are tagged with natural language-based tags like noun, adjective, pronoun, etc. POS tagging is the process of marking up a word in a corpus to a corresponding part of a speech tag, based on its context and definition. To understand the meaning of any sentence or to extract relationships and build a knowledge graph, POS tagging is a very important step.

There are different techniques for POS tagging:

- Lexical-based methods – This method assigns the POS tag the most frequently occurring with a word in the training corpus.
- Rule-based methods – Assigns POS tags based on rules. For example, we can have a rule that says words ending with 'ed' or 'ing' must be assigned to a verb. Rule-based techniques can be used along with lexical-based approaches to allow POS tagging of words that are not present in the training corpus but are there in the testing data.
- Probabilistic methods – This method assigns the POS tags based on the probability of a particular tag sequence occurring. Conditional Random Fields (CRFs) and Hidden Markov Models (HMMs) are probabilistic approaches to assigning a POS tag.
- Deep learning methods – Recurrent neural networks can also be used for POS tagging.

The list of POS tags, with examples of what each POS stands for, is as follows:

CC – coordinating conjunction
CD – cardinal digit
DT – determiner
EX – existential there (e.g. 'there is' … think of it like 'there exists')

FW – foreign word
IN – preposition/subordinating conjunction
JJ – adjective 'big'
JJR – adjective, comparative 'bigger'
JJS – adjective, superlative 'biggest'
LS – list marker 1
MD – modal could, will
NN – noun, singular 'desk'
NNS – noun, plural 'desks'
NNP – proper noun, singular 'Harrison'
NNPS – proper noun, plural 'Americans'
PDT – predeterminer 'all the kids'
POS – possessive ending parent's
PRP – personal pronoun I, he, she
PRP$ – possessive pronoun my, his, hers
RB – adverb very, silently,
RBR – adverb, comparative better
RBS – adverb, superlative best
RP – particle give up
UH – interjection
VB – verb, base form take
VBD – verb, past tense took
VBG – verb, gerund/present participle taking
VBN – verb, past participle taken
VBP – verb, singular, present, take
VBZ – verb, third person singular, present takes
WDT – wh-determiner which
WP – wh-pronoun who, what
WP$ – possessive wh-pronoun whose
WRB – wh-abverb where, when

Example: Can you please buy me an Arizona Ice Tea? It's $0.99.
POS tagged Output:

[('Can', 'MD'), ('you', 'PRP'), ('please', 'VB'), ('buy', 'VB'), ('me', 'PRP'), ('an', 'DT'), ('Arizona', 'NNP'), ('Ice', 'NNP'), ('Tea', 'NNP'), ('?', '.'), ('It', 'PRP'), (' 's', 'VBZ'), ('$', '$'), ('0.99', 'CD'), ('.', '.')

This is an example showing a sentence tagged with POS tags.
Applications of POS tagging:
POS tagging finds applications in

- Named entity recognition (NER)
- Sentiment analysis
- Question answering
- Word sense disambiguation, and so on

1.6 EMERGING TECHNIQUES FOR SUMMARIZATION

In the field of text summarization, some of the evolving techniques like machine learning, sentiment-based summarization, parallel algorithms, and optimization are discussed briefly in this section.

Extractive text summarization methods based on machine learning can be broadly classified as unsupervised learning and supervised learning methods. Recent works rely on unsupervised learning methods for text summarization.

Supervised learning methods are based on standard algorithms like naïve Bayes and neural networks. However, this method requires summarization problem to be modeled either as classification or some other standard problems. Unsupervised techniques use concept-based, graphical, or fuzzy logic-based methods for summarization.

The main challenge for summarization is the evaluation of summaries (either automatically or manually). The main problem in evaluation comes from the impossibility of building a standard against which the results of the systems that have to be compared. Since the explosion of data becomes exponential, there is a key requirement in developing large-scale summarization algorithms based on machine learning. This has been explored in Chapter 2.

Sentiment Analysis-Based Summarization

Sentiment-based summarization specifically focuses on summarizing the text with sentiments or opinions. In recent years, opinion mining or sentiment analysis has been an active research area in text mining and analysis, natural language processing. Aspect-based summarization is a fine-grained analysis which summarizes sentiments based on significant features or aspects from text. It produces a short textual summary based on the important aspects and its opinions. Yadav and Chatterjee (2016) studied the use of sentiments in text summarization. This is considered as a separate domain by itself. There are two major techniques used for aspect based summarization. They are rule-based and machine learning methods. They are explained in detail in Chapter 3.

Other challenges that are facing opinion summarization, especially in commercial services include poor accuracy resulting from difficulties in understanding language along with the scalability issue which requires deeper and complicated NLP technologies to be handled.

Parallel Processing Approaches

Parallel processing approaches have been studied extensively in the past in the literature for text summarization tasks. There are different kinds of algorithms based on parallel processing like clustering, genetic algorithm, parallel topic modeling, etc. In recent days, evolutionary approaches and parallel algorithms using large-scale tools are studied because of their efficiency in handling large-scale text data. Chapter 4 elaborates on these techniques and their limitations.

Optimization Approaches

Optimization is an active research field for all problems. Evolutionary algorithms and sentence ranking–based approaches play a crucial role in optimization of text summarization task. Current research includes developing parallel processing algorithms based on MapReduce and other paradigms. Chapter 5 exclusively deals with several optimization approaches for summarization and their challenges.

Cognition-Based Approaches

The potential of combining emotion cognition with deep learning in sentiment analysis of social media data for summarization has been explored by Wu et al. (2020). Combining emotion-based approaches had yielded better results in the field of summarization. There are other works that simulate cognitive processes such as read-again techniques used by humans for encoding sequence-to-sequence models with neural networks by Shi et al. (2018).

1.7 SCOPE OF THE BOOK

So far roughly around ten books have been published about automatic text summarization: Much of these are concerned with automatic summarization. This book is aimed at people who are interested in automatic summarization algorithms: researchers, undergraduate and postgraduate students in NLP, PhD students, computer scientists, engineers, mathematicians, and linguists. This book aims to provide an exhaustive and current method used for automatic text summarization. It will offer an elaboration of all stages of summarization starting from preprocessing. The readers will be able to increase their knowledge of the subject as well as research perspective.

The book is divided into seven chapters:

- Chapter 1. Introduction
- Chapter 2. Large-Scale Summarization Using Machine Learning Approach
- Chapter 3. Sentiment Analysis Approach to Text Summarization
- Chapter 4. Text Summarization Using Parallel Processing Approach
- Chapter 5. Optimization Approaches for Text Summarization
- Chapter 6. Performance Evaluation of Large-Scale Summarization Systems
- Chapter 7. Applications and Future Directions

This book gives first and foremost a realistic look at text summarization and its applications. A coherent overview of the field will be given along with all the current state-of-the-art techniques and future research directions.

Exercises on TF-IDF, Vector Space Model for Text Representation

1. Consider the following three example sentences:
   'I like to play football'
   'Did you go outside to play tennis'
   'John and I play tennis'
   Represent them using BoW model and compute TF_IDF also.

2. Consider the table of term frequencies for three documents denoted Doc1, Doc2, and Doc3 in Table 1.5. Compute the tf-idf weights for the terms car, auto, insurance, and best, for each document, using the idf values from Table 1.6.
3. Given a document with the terms A, B, and C with the following frequencies A: 3, B: 2, and C: 1. The document belongs to a collection of 10,000 documents. The document frequencies are: A: 50, B: 1300, and C: 250. Compute the normalized tf and tf-idf and compare them.
   The idf values are: A idf = log(10,000/50) = 5.3; B idf = log(10,000/1300) = 2.0; C idf = log(10,000/250) = 3.7.
4. Assume we have the following set of five documents (only terms that have been selected are shown):
   Doc1: cat cat cat
   Doc2: cat cat cat dog
   Doc3: cat dog mouse
   Doc4: cat cat dog dog dog
   Doc5: mouse
   (i) How would you represent each document as a vector?
   (ii) Calculate the TF-IDF weights for the terms.
5. Consider the following collection of five documents:
   Doc1: we wish efficiency in the implementation of a particular application

TABLE 1.5
Term Frequencies

	Doc1	Doc2	Doc3
Car	27	4	24
Auto	3	33	0
Insurance	0	33	29
Best	14	0	17

TABLE 1.6
IDF Values

	Da_{ft}	Def_t
Car	18,165	1.65
Auto	6723	2.08
Insurance	19,241	1.62
Best	25,235	1.5

Note: Here IDF of terms with various frequencies are given using Reuter's collection of 806,791 documents.

Doc2: the classification methods are an application of Li's ideas
Doc3: the classification has not followed any implementation pattern
Doc4: we have to take care of the implementation time and implementation efficiency
Doc5: the efficiency is in terms of implementation methods and application methods

Assuming that every word with six or more letters is a term and that terms are ordered in the order of appearance:

1. Give the representation of each document in the Boolean model.
2. Give the representation in the vector model using tf-idf weights of documents Doc1 and Doc5.

6. Compare term frequency and inverse document frequency.
7. Below is a snippet of the first few lines of text from the book *A Tale of Two Cities* by Charles Dickens, taken from Project Gutenberg.
   It was the best of times,
   it was the worst of times,
   it was the age of wisdom,
   it was the age of foolishness,

   Create a bag-of-words model for the above lines. Treat each line as a separate 'document' and the four lines as the entire corpus of documents.
8. Given a document with term frequencies: A(3), B(2), and C(1). Assume the collection contains 10,000 documents, and document frequencies of these terms are: A(50), B(1300), and C(250).
   Document A: 'A dog and a cat.'
   Document B: 'A frog.'

   Consider the two documents A and B. Find the vocabulary and sort. Represent the two documents as vectors.
9. Use the following four documents:
   D1: The cat sat on the mat. The cat was white.
   D2: The brown coloured dog was barking loudly at the cat.
   D3: The mat was green in colour.
   D4: The dog pulled the mat with his teeth. The cat still sat on the mat.
   Represent them using BoW model and vector space model.

REFERENCES

Shi, T., Keneshloo, Y., Ramakrishnan, N., and Reddy, C. K. 2018. "Neural abstractive text summarization with sequence-to-sequence models." *arXiv preprint arXiv*:1812.02303.

Wu, H. C., Luk, R. W. P., Wong, K. F., and Kwok, K. L. 2008. "Interpreting TF-IDF term weights as making relevance decisions." *ACM Transactions on Information Systems* 26(3): 1. doi: 10.1145/1361684.1361686.

Wu, P., Li, X., Shen, S., and He, D. 2020. "Social media opinion summarization using emotion cognition and convolutional neural networks." *International Journal of Information Management* 51: 101978.

Yadav, N., and Chatterjee, N. 2016. "Text summarization using sentiment analysis for DUC data." In *2016 International Conference on Information Technology (ICIT)* (pp. 229–234). Bhubaneswar. doi: 10.1109/ICIT.2016.054.

SAMPLE CODE

//DOCUMENT REPRESENTATION

```
import java.io.BufferedWriter;
import java.io.File;
import java.io.FileInputStream;
import java.io.FileWriter;
import java.sql.ResultSet;
import java.sql.SQLException;
import java.util.ArrayList;
import java.util.NavigableMap;
import java.util.Scanner;
import java.util.TreeMap;
import javax.swing.table.DefaultTableModel;
public class BOWords extends javax.swing.JFrame { double
       dmean=0;
public BOWords(double d) { dmean=d;
initComponents(); }
DBConnection db = new DBConnection();
private void jButton1ActionPerformed(java.awt.event.Action
        Event evt) {
StringBuilder s = new StringBuilder(); int c; String
              words[];
int column=2; try {
ResultSet rs=db.execute("select * from tbl_tf where tf>" +
dmean + " order by tf desc"); while(rs.next()){
s.append(rs.getString(1)+" "); }} catch
        (SQLException ex) {
Logger.getLogger(BOWords.class.getName()).log(Level.SEVERE,
null, ex); } words =
       s.toString().replaceAll("\n"," ").split(" ");
NavigableMap<String, Integer> nmap = new TreeMap<String,
Integer>(); for (int i = 0; i <
       words.length; i++) {
words[i] = words[i].toLowerCase().trim();
       nmap.put(words[i], 1); }
nmap = nmap.descendingMap(); int
       commaCorrection = 0;
```

```
for (String key : nmap.keySet()) {
       if(commaCorrection == 0 ) {
       jTextArea1.append(key+"\n");
       System.out.print(key);
       commaCorrection =1; } else {
       System.out.print(","+key);
       jTextArea1.append(key+"\n"); } }
```

//CALCULATION OF TF-IDF, WEIGHT

```
import java.io.File;
import java.io.FileNotFoundException; import
       java.io.FileOutputStream; import
       java.io.PrintStream;
import java.util.ArrayList; import
       java.util.List; import
       java.util.Scanner;
public class tfidf extends javax.swing.JFrame {
private void jButton1ActionPerformed(java.awt.event.Action
Event evt) { try{
File folder = new File("D:\\PSOSummarizationuptoPSO\\hotel
s\\beijing"); File[] listOfFiles =
       folder.listFiles();
String fi[]=new String[listOfFiles.length]; for (int i =
       0; i <listOfFiles.length; i++) { if
       (listOfFiles[i].isFile()) {
       fi[i]=listOfFiles[i].getName(); } }
List<List<String>> documents = new ArrayList<>();
ArrayList<String>[]
       doc = new ArrayList[listOfFiles.length]; for(int
       k=0;k<listOfFiles.length;k++) {
File f=new File("D:\\PSOSummarizationuptoPSO\\hotels\\be
ijing\\"+fi[k]); Scanner s=new
       Scanner(f);
doc[k]=new ArrayList<>();
while(s.hasNext()) {
       doc[k].add(s.next()); }
       documents.add(doc[k]); }
       tfidf calculator = new tfidf();
       double tfidf1 ;
for(int i=0;i<documents.size();i++) {
FileOutputStream fout=new
       FileOutputStream("D:\\PSOSummarizationuptoPSO\\hot
           els\\beijing\\"+i);
PrintStream p=new PrintStream(fout); for(int
       j=0;j<documents.get(i).size();j++) {
tfidf1 = calculator.tfIdf(documents.get(i), documents
,documents.get(i).get(j) ); p.print();
```

```
p.println(documents.get(i).get(j)+"="+tfidf1);
       jTextArea1.append(documents.get(i).get(j)+"="+tfidf1);
       jTextArea1.append("\n");           } }
FileOutputStream fout=new FileOutputStream("C:\\Users\\S
harmi\\Desktop\\prodataset\\op\\tf\\"+fi[k]) }
catch(Exception e){} }

public double tf(List<String> doc, String term) { double
result = 0;

for (String word : doc) {

if (term.equalsIgnoreCase(word))
result++; } return result / doc.size();}

public double idf(List<List<String>> docs, String term) {
double n = 0;
for (List<String> doc : docs) { for
       (String word : doc) {
if (term.equalsIgnoreCase(word)) { n++;
       break; } } }return
       Math.log(docs.size() / n);}
public double tfIdf(List<String> doc, List<List<String>>
docs, String term) { return tf(doc,
       term) * idf(docs, term);}
public tfidf() { initComponents(); }
```

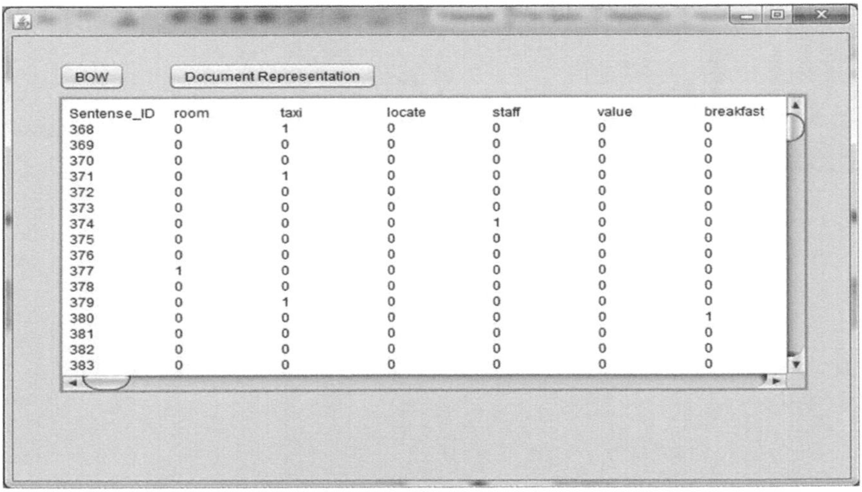

BOW   Document Representation

Sentense_ID	room	taxi	locate	staff	value	breakfast
368	0	1	0	0	0	0
369	0	0	0	0	0	0
370	0	0	0	0	0	0
371	0	1	0	0	0	0
372	0	0	0	0	0	0
373	0	0	0	0	0	0
374	0	0	0	1	0	0
375	0	0	0	0	0	0
376	0	0	0	0	0	0
377	1	0	0	0	0	0
378	0	0	0	0	0	0
379	0	1	0	0	0	0
380	0	0	0	0	0	1
381	0	0	0	0	0	0
382	0	0	0	0	0	0
383	0	0	0	0	0	0

FIGURE 1.2 Document representation – BoW model

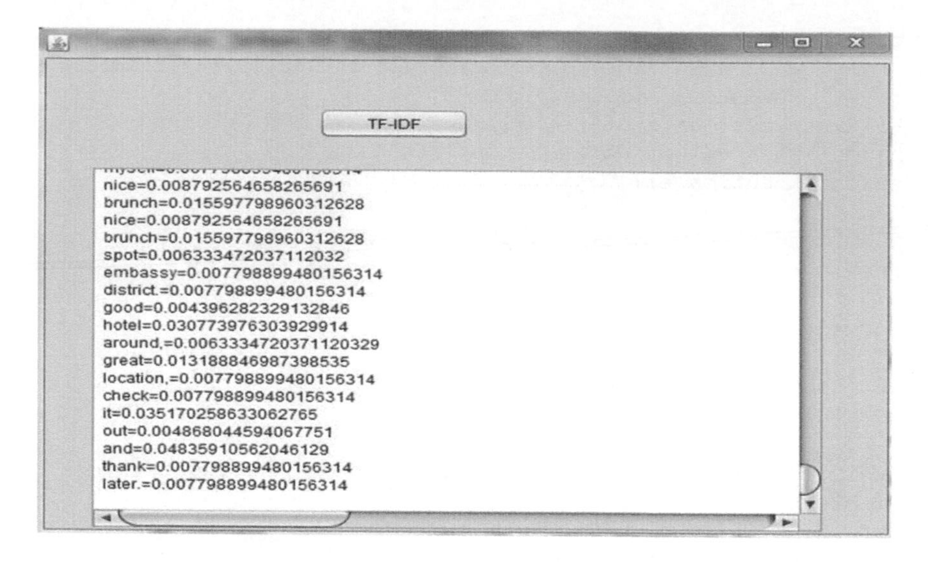

FIGURE 1.3 Document representation – TF-IDF model

Sample Screenshots

The sample screenshots for the above given sample code – Document representation for the BoW model is represented in Figure 1.2, and document representation using TF-IDF model is represented in Figure 1.3.

2 Large-Scale Summarization Using Machine Learning Approach

2.1 SCALING TO SUMMARIZE LARGE TEXT

The need for automatic summarization systems are:

- It creates summaries that reduce users' reading time.
- When researching documents, the generated summaries make the selection process easier.
- It improves the effectiveness of indexing.
- Automatic summarization system is less biased than human summarizers.
- The produced summaries are also useful in question-answering systems.
- Using automatic summarization system enables commercial abstract services to increase the number of texts they are able to process.

Most summarization techniques require the data to be centralized, which may not be feasible in many cases due to computational and storage limitations. In today's world, data growth is exponential. The huge increase of data emerging by the progress of technology and various sources makes automatic text summarization of large scale of data a challenging task. This necessitates the requirement of large-scale algorithms, where traditional algorithms are not suitable for processing voluminous text.

2.2 MACHINE LEARNING APPROACHES

Machine learning methods can be broadly classified into two categories: supervised and unsupervised methods. Supervised methods learn from the training samples for which the model is going to be used. Unsupervised methods do not require any training. They simply follow the model/algorithm and generate the required output. In text summarization problem, machine learning models [2] are usually trained to understand documents and extract useful information before creating the required summarized texts.

DOI: 10.1201/9781003371199-2

2.2.1 Different Approaches for Modeling Text Summarization Problem

There are many methods based on supervised and unsupervised learning for text summarization problem. Most of them are based on classification. Few works are based on clustering and neural networks. Recent trends focus on utilization of deep learning methods for summarization.

2.2.2 Classification as Text Summarization

Classification is a supervised learning technique which has been used for solving many real-world problems. In text summarization problems, it has been used widely with different models. In general classification implies arrangement in groups or categories according to the features and similarities established between one reference sample and test input samples. In summarization the sentences or documents to be summarized are grouped into one or more categories. Here sentence scores or significance score for the sentence to be included in the final summary acts as the criteria. There are some other criteria's which are also used for two-class or multi-class problem for summarization. The phases of summarization when modeled as classification are elaborated in the following sections.

2.2.2.1 Data Representation

Data representation plays an important role in the classification task. Unstructured data, particularly free-running text data, has to be transformed into a structured data. This structured data should be transformed into an effective document representation model to build an efficient classification system. Some of the prominent methods found in the literature such as bag-of-words (BoW), vector space model, and term-weighting schemes are discussed in Chapter 1. The methods which are found suitable for text data classification task are detailed in this section.

Ontology-based representation, n-gram model, Universal Networking Language (UNL), and word embedding model are the widespread schemes available. Ontology denotes a set of concepts defined and linked to a domain. It can be seen as a knowledge base representation for a specific domain, consisting of definitions and relationships of concepts. This representation preserves the semantic relationships of terms or words in the document. The authors, Shah et al. [15], provided ontology-based representation of text documents. In the representation, ontology is defined as a set consisting of lexicon, concepts, reference function, and a hierarchy for the concepts. A hierarchy of concepts is a kind of "taxonomy" where each term may have multiple children and multiple parents. An example of hotel domain is shown below:

Lexicon, L = {Hotel, Grand Hotel, Hotel Schwarzer Adler, Accommodation,)
Concepts C* = {ROOT, HOTEL, ACCOMMODATION …}
Reference function, F = {(Hotel, HOTEL), (Grand Hotel, HOTEL), (Hotel Schwarzer Adler, HOTEL) …}, i.e. "Hotel", "Grand Hotel" and "Hotel Schwarzer Adler" refer to the concept HOTEL.

Hierarchy, H = {(HOTEL, ACCOMMODATION), (ACCOMMODATION, ROOT) …}

The next standard scheme is n-grams, which are extracted from a long string in a document. This scheme is used for preprocessing by Canvas. An example for n-gram representation of the word TEXT is:

Bigrams: _T, TE, EX, XT, T_
Trigrams: _TE, TEX, EXT, XT_
Quadgrams: _TEX, TEXT, EXT_

This representation is found useful in many applications like information retrieval, clustering, summarization, etc. In an n-gram scheme, it is very difficult to decide the number of grams to be considered for effective document representation.

Choudhary and Bhattacharyya proposed a Universal Networking Language (UNL) to represent a document. The UNL represents the document in the form of a graph. The graph contains words as nodes and relation between them as links.

Figure 2.1 shows an example graph using UNL for the sentence 'John, who is the chairman of the company, has arranged a meeting at his residence.'

Words in the document are: John, chairman, company, meeting, residence

Relations are: place, position, agent, object, modifier, and adjective

This method requires the construction of a graph for every document. So this is suitable only for small document collection. These are some of the text representations used for the classification task in machine learning.

Examples of unigrams, bigrams, and trigrams:

Sentence = 'You will face many defeats in life, but never let yourself be defeated.'
Unigrams
('You',)
('will',)

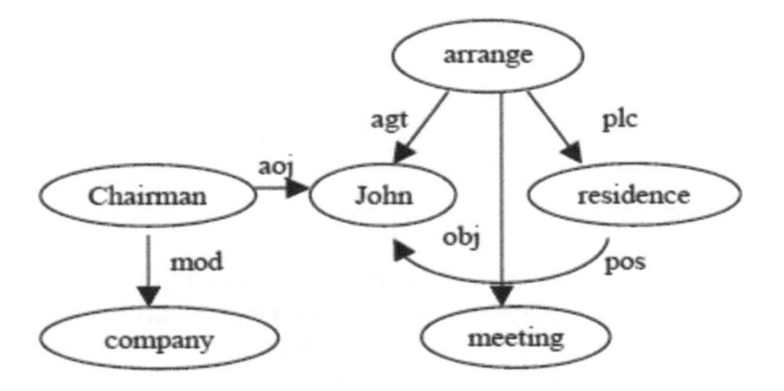

FIGURE 2.1 Example of Universal Networking Language graph

('face',)
('many',)
('defeats',)
('in',)
('life,',)
('but',)
('never',)
('let',)
('yourself',)
('be',)
('defeated.',)
sentence = 'The purpose of our life is to happy'

Bigrams

('The', 'purpose')
('purpose', 'of')
('of', 'our')
('our', 'life')
('life', 'is')
('is', 'to')
('to', 'happy')

Sentence = 'Whoever is happy will make others happy too'

Trigrams

('Whoever', 'is', 'happy')
('is', 'happy', 'will')
('happy', 'will', 'make')
('will', 'make', 'others')
('make', 'others', 'happy')
('others', 'happy', 'too')

Example: UNL Expression

An example of a UNL expression of a sentence is as follows:

'I can hear a dog barking outside'
{unl}
aoj(hear(icl>perceive(agt>thing,obj>thing)).@entry.@ability, I)
obj(hear(icl>perceive(agt>thing,obj>thing)).@entry.@ability, :01)
agt:01(bark(agt>dog).@entry, dog(icl>mammal))
plc:01(bark(agt>dog).@entry, outside(icl>place))
{/unl}

In the above UNL expression, aoj, agt, and obj are the relation labels; I, bark(agt>dog), dog(icl>mammal), hear(icl>perceive(agt>thing,obj>thing)), and outside(icl>place) are the universal words; and :01 appears three times in the example.

2.2.2.2 Text Feature Extraction

Feature extraction is considered to be essential in classification, since the overall performance is based on the features extracted from the documents. Feature extraction is the process of extracting a key subset of features from the data for improving the classification task. Identification of the related features in the text is important for improving performance. Extraction of text features is the process of taking out a list of words from the text data and then transforming them into a feature set. Some of the prominent methods for feature extraction are principal component analysis (PCA), neural networks, etc. Both supervised and unsupervised learning for extracting essential features from the text dataset are discussed.

Principal component analysis is one of the best methods for feature extraction. It is a method used for the linear transformation of feature vector. It is a statistical method that uses an orthogonal linear transformation to transform a high-dimensional feature space into a new low-dimensional feature subspace. The number of transformed principal components may be less than or equal to the original variables. A detailed mathematical process is explained in the tutorial by Shlens (2005). PCA has advantages like computationally inexpensive, handle sparse and skewed data, and the ability to remove noisy features.

Back propagation network is one of the techniques under artificial neural networks. In this method, pairs of input and reference patterns are presented and evaluated using weights. These weights are then progressively moved towards the output layer. Error rates are computed after the first iteration to adjust the weights in order to minimize the loss. Repeated training for all the samples will yield a model with better accuracy to get closer to the minimum loss/cost at every step.

Convolutional neural networks (CNNs) are networks composed of several layers of convolutions with nonlinear activation functions applied to the results. Traditional layers are fully connected, whereas CNNs use local connections. Each layer applies different filters as much as thousand filters and combines their results. The limitation of CNNs is that they require a huge amount of time and space which makes it difficult to process huge volumes of data. This necessitates the deep learning–based neural network model to be used for feature extraction. Several text-based features that are used for summarization based on sentence score classification in recent works are listed below.

Title Feature

The sentence that contains the word(s) in the document title is given a high score. This can be computed by counting the number of matching characters between the words in a sentence and the words in the document title as shown in Equation 2.1.

$$f_1 = \frac{\text{Number of title words in sentence}}{\text{Number of words in document title}} \quad (2.1)$$

Sentence to Sentence Similarity

This feature measures the similarity between sentence S and other sentences in the document set. The similarity is computed by using Jiang's semantic similarity measure as shown in Equation 2.2.

$$f_2 = \frac{\sum sim(S_i, S_j)}{\max\left(\sum sim(S_i, S_j)\right)} \quad (2.2)$$

Sentence Location

The placement of the sentence in the text is important, where the beginning sentences in the document are given high scores as per Equation 2.3.

$$f_3 = \frac{\text{Length of the document} - \text{sentence position} + 1}{\text{Length of the document}} \quad (2.3)$$

Numerical Data

Sentence containing statistical data is important for summary, where the feature can be computed by using Equation 2.4.

$$f_4 = \frac{\text{Number of numerical data in the sentence}}{\text{Length of the sentence}} \quad (2.4)$$

Temporal Feature

Temporal extraction is used to extract explicit event containing date or time expressions in the article and is measured based on Equation 2.5.

$$f_5 = \frac{\text{Number of temporal information in the sentence}}{\text{Length of the sentence}} \quad (2.5)$$

Length of Sentence

Long sentence is considered to inherit important information. The length of the sentence is computed by using Equation 2.6.

$$f_6 = \frac{\text{Number of words in the sentence}}{\text{Number of words in the longest sentence}} \quad (2.6)$$

Proper Noun

A sentence containing proper noun is considered to be an important sentence. The scores of such sentence are computed with Equation 2.7.

$$f_7 = \frac{\text{Number of proper nouns in the sentence}}{\text{Length of the sentence}} \quad (2.7)$$

Number of Nouns and Verbs

Sentences that contain more nouns and verbs are considered to inherit important information. The score can be computed with Equation 2.8.

$$f_8 = \frac{\text{Number of nouns and verbs in the sentence}}{\text{Length of the sentence}} \tag{2.8}$$

Frequent Semantic Term

This feature is used to determine the commonness of a term. A term that is used frequently is probably related to the topic of the document. The top 10 words are considered as the maximum number of frequent semantic terms. The computation is shown in Equation 2.9.

$$f_9 = \frac{\text{Number of frequent terms in the sentence}}{\text{Max}\left(\text{number of frequent terms}\right)} \tag{2.9}$$

These are some of the text and sentence based features used for sentence scoring and summarization.

All the above features capture some of the significant aspects for summarization, such as nouns, verbs, number of frequent terms, number of title words, and similarity between sentences. All these features are used for classifying sentences as summary or non-summary sentences.

2.2.2.3 Classification Techniques

Classification has been widely adopted in the literature for text summarization. There are standard algorithms such as K-nearest neighbor (KNN), Naïve Bayes, and sentence scoring–based approaches which model summarization as classification problem.

K-nearest neighbor is a supervised learning method. Jo (2019) proposed an approach based on string vectors. In this supervised learning approach, the training texts are encoded into string vectors. A novice text is encoded into a string vector; its similarities with ones which represent the training ones are computed by Equation 2.10, which was proposed by Sarkar (2009).

$$Sim\left(S_i, S_j\right) = \frac{\left(2 * \left|S_i \cap S_j\right|\right)}{\left|S_i\right| + \left|S_j\right|} \tag{2.10}$$

where S_i and S_j are any two sentences belonging to the input collection of sentences. The numerator $|S_i \cap S_j|$ represents the number of matching words between two sentences, and $|S_i|$ is the length of the ith sentence, where length of a sentence = number of words in the sentence.

The most K similar training texts are selected as its nearest neighbors. The label of the novice text is decided by voting one of the nearest neighbors. This method is applied to news articles, which are processed as described below.

The sample paragraphs (training sets) which are labeled with summary or non-summary are gathered domain by domain from news articles. They are encoded into string vectors. The text which is assumed to be tagged with its own domain is partitioned into a list of paragraphs. For each paragraph, its similarities are computed with the sample ones in the corresponding domain. For each paragraph, its K-nearest sample ones are selected, and its label is decided by voting their labels. The ones which are classified with label 'summary' are extracted as the text summary. This approach yields good classification accuracy in four domains of news articles.

The method used by Shah et al. incorporates latent semantic analysis (LSA) for choosing representative sentences. Then Naïve Bayes classifier is trained as a model for predicting the summary sentences. The prediction is built upon singular value decomposition, which uses feature extraction and recursive feature elimination as the key concepts. Ramanujam et al. showed that Naïve Bayes can be successfully used for multi-document summarization also. In this work, time-stamp-based approach with Naïve Bayesian Classification was used. In multi-document summarization, input is received from multiple sources. Keywords are given in the training model, and these are used to extract frequent document sentences. These sentences are scored by using positional value score and term frequency/inverse document frequency (TF-IDF) score. The visiting times for each document and also selection of the document for processing are computed with time stamp. These sentences are arranged based on scores, and high-scored sentences are used for summary generation.

Support vector machines are eminently used for classification and have been found to perform better in summarization task. This has been applied to Telugu text documents. In this dimensionality, problem in classification is overcome by using the features such as sentence location, term feature weighting, and centrality. This has been applied to Telugu newspaper articles and found to have a high F_1 measure.

Finally classification data have been used to improve the performance of text summarization system for multi-document summarization by Ziqiang et al. In this method, a convolutional neural network is trained for projecting a document into distributed representation. Then a softmax classifier predicts the category of the document. Now, the summarization model shares the same projection to generate document embeddings, which is transformed into summary embeddings. These embeddings are matched with the meaning of the reference summaries to the maximum level. The model is trained on 2 years of Document Understanding Conference (DUC) data and achieves better performance compared to other baseline methods. There are many supervised learning-based classification performed well for summarization. Likewise unsupervised learning methods like clustering also perform remarkably in summarization task.

Practical Example of K-Nearest Neighbor Algorithm

Step-by-step explanation of how to compute K-nearest neighbors (KNN) algorithm:

1. Determine parameter K = number of nearest neighbors.
2. Calculate the distance between the query instance and all the training samples.

3. Sort the distance and determine the nearest neighbors based on the K-th minimum distance.
4. Gather the category of the nearest neighbors.
5. Use the simple majority of the category of nearest neighbors as the prediction value of the query instance.

Given: Table 2.1, which specifies the values of X_1 and X_2 and classification.

Now the factory produces a new paper tissue that passes laboratory test with $X_1 = 3$ and $X_2 = 7$. Without another expensive survey, guess what the classification of this new tissue is?

1. Determine parameter K = number of the nearest neighbors. Suppose use $K = 3$.
2. Calculate the distance between the query instance and all the training samples.

Coordinate of query instance is (3, 7); instead of calculating the distance, we compute square distance which is faster to calculate (without square root) as in Table 2.2.

3. Sort the distance and determine the nearest neighbors based on the K-th minimum distance as in Table 2.3.
4. Gather the category of the nearest neighbors as mentioned in Table 2.4. Note in the second row last column that the category of nearest neighbor (Y) is not included because the rank of this data is more than 3 (=K).

TABLE 2.1
Values of X_1 and X_2 and Classification

X_1 = Acid Durability (s)	X_2 = Strength (kg/m^2)	Y = Classification
7	7	Bad
7	4	Bad
3	4	Good
1	4	Good

TABLE 2.2
Calculation of Square Distance to Query Instance (3, 7)

X_1 = Acid Durability (s)	X_2 = Strength (kg/m^2)	Square Distance to Query Instance (3, 7)
7	7	$(7 - 3)^2 + (7 - 7)^2 = 16$
7	4	$(7 - 3)^2+(4 - 7)^2 = 25$
3	4	$(3 - 3)^2+(4 - 7)^2 = 9$
1	4	$(1 - 3)^2+(4 - 7)^2 = 13$

TABLE 2.3
Determining Whether the Sample Is Included in 3-Nearest Neighbor

X_1 = Acid Durability (s)	X_2 = Strength (kg/m²)	Square Distance to Query Instance (3, 7)	Rank Minimum Distance	Is It included in 3-Nearest Neighbors?
7	7	$(7-3)^2+(7-7)^2 = 16$	3	Yes
7	4	$(7-3)^2+(4-7)^2 = 25$	4	No
3	4	$(3-3)^2+(4-7)^2 = 9$	1	Yes
1	4	$(1-3)^2+(4-7)^2 = 13$	2	Yes

TABLE 2.4
Category Based on Nearest Neighbor

X_1 = Acid Durability (s)	X_2 = Strength (kg/m²)	Square Distance to Query Instance (3, 7)	Rank Minimum Distance	Is It Included in 3-Nearest Neighbors?	Y = Category of Nearest Neighbor
7	7	$(7-3)^2+(7-7)^2 = 16$	3	Yes	Bad
7	**4**	$\mathbf{(7-3)^2+(4-7)^2 = 25}$	**4**	**No**	**–**
3	4	$(3-3)^2+(4-7)^2 = 9$	1	Yes	Good
1	4	$(1-3)^2+(4-7)^2 = 13$	2	yes	Good

Bold indicates outlier data.

5. Use the simple majority of the category of nearest neighbors as the prediction value of the query instance.

We have two good and one bad; since 2 > 1, then we conclude that a new paper tissue that passes laboratory test with **$X_1 = 3$ and $X_2 = 7$ is included in Good category**.

2.2.3 Clustering as Text Summarization

Unsupervised learning does not contain any training data. It learns from the parameters or the algorithmic techniques involved. K-means clustering, a fast algorithm, has been used by Prathima and Divakar (2018). This work groups the sentences in the documents and they are clustered based on their TF-IDF values and similarity with other sentences. Then Support Vector Machine (SVM) is used as a cascaded classifier for summary generation. Additionally in this work, for the transmission of documents Advanced Encryption Standard algorithm (AES) is used. The authors demonstrated better performance using a query-based retrieval application. Multi-document summarization task has been the focus of many researchers to adapt clustering.

Multi-document summarization is the task of summarizing documents from multiple sources. Sentence clustering has been widely used for this task. Sarkar (2009) attempted to utilize this concept for summarization. The clustering-based multi-document summarization performance largely depends on three important factors: (i) clustering sentences, (ii) cluster ordering, and (iii) selection of representative sentences from the clusters. Clustering of sentences was performed using incremental dynamic method using similarity histogram. Ordering is done based on the cluster importance, which is computed by the sum of the weights of the content words of a cluster. Representative sentences are selected using three different methods:

- Arbitrary (random) selection
- Longest candidate selection
- Sentence selection based on its similarity to the centroid of the input document set

The system was evaluated on the DUC dataset and was found competitive with other multi-document summarization systems. Kogilavani and Balasubramani (2010) used a document clustering approach based on feature profiling. Related documents are grouped into the same cluster using document clustering algorithm. Feature profile is generated by considering word weight, sentence position, sentence length, sentence centrality, and proper nouns in the sentence and numerical data in the sentence.

Feature profile–based sentence score is calculated for each sentence. Cluster-wise summary is generated. The system is evaluated using precision, recall, and f-measure on news genre articles showing remarkable performance. In Siddiki and Gupta's (2012) approach, multi-document summarization was generated from single-document summaries created. Single-document summary is generated using document feature, sentence reference index feature, location feature, and concept similarity feature. Sentences from single-document summaries are clustered, and top most sentences from each cluster are used for creating multi-document summary. The next method used by Anjali et al. combined both document and sentence clustering for query-based multi-document summarization. In this approach, query-dependent corpus is created, and features like noun phrase, sentence length, and cue phrases are utilized for sentence scoring. Documents are clustered using cosine similarity to create every document cluster, and sentences are clustered based on their similarity values. Score all sentences, and the best-scored sentences are picked up to create summary. Clustering has been used majorly for multi-document summarization. The following section is discussed with the latest deep learning–based approach for summarization.

A Step-by-Step Approach for Clustering

In this illustration we are going to cluster Wikipedia articles using k-means algorithm. The steps are as follows:

1. Fetch some Wikipedia articles.
2. Represent each article as a vector.

3. Perform k-means clustering.
4. Evaluate the result.

For all the steps, sample code is given in Python.

1. Fetch Wikipedia articles

Using the Wikipedia package, it is very easy to download the content from Wikipedia. For this example, we will use the content of the articles for:

Data science
Artificial intelligence (AI)
Machine learning
European Central Bank
Bank
Financial technology
International Monetary Fund
Basketball
Swimming
Tennis

The content of each Wikipedia article is stored in wiki_list while the title of each article is stored in variable title.

Sample code

```
articles=['DataScience','Artificialintelligence','MachineLearning','European
   Central Bank','Bank','Financial technology','International Monetary Fun
   d','Basketball','Swimming','Tennis']
wiki_lst=[]
title=[]
for article in articles:
print("loading content: ",article)
wiki_lst.append(wikipedia.page(article).content)
title.append(article)
```

2. Represent each article as a vector

Since we are going to use k-means, we need to represent each article as a numeric vector. A popular method is to use TF-IDF. Put it simply, with this method, for each word w and document d, we calculate:

tf(w,d): the ratio of the number of appearances of w in d divided by the total number of words in d.
idf(w): the logarithm of the fraction of the total number of documents divided by the number of documents that contain w.
$\text{tfidf}(w,d) = \text{tf}(w,d) \times \text{idf}(w)$

It is recommended that common, stop words are excluded.

3. Perform k-means clustering

Each row of variable X is a vector representation of a Wikipedia article. Hence, we can use X as input for the k-means algorithm.

Optimal number of $K = 4$ or 6 (obtained using elbow method).

Sample Code

```
model = KMeans(n_clusters=true_k, init='k-means++', max_iter=200, n_init=10)
model.fit(X)
labels=model.labels_
wiki_cl=pd.DataFrame(list(zip(title,labels)),columns=['title','cluster'])
print(wiki_cl.sort_values(by=['cluster']))
```

Result of Clustering

The result of clustering is shown in Figure 2.2.

4. Evaluate the result

Since we have used only ten articles, it is fairly easy to evaluate the clustering just by examining what articles are contained in each cluster.

Sample Code

```
from wordcloud import WordCloud
result={'cluster':labels,'wiki':wiki_lst}
result=pd.DataFrame(result)
for k in range(0,true_k):
  s=result[result.cluster==k]
text=s['wiki'].str.cat(sep=' ')
text=text.lower()
text=' '.join([word for word in text.split()])
wordcloud = WordCloud(max_font_size=50, max_words=100,
background_color="white").generate(text)
print('Cluster: {}'.format(k))
print('Titles')
```

	title	cluster
3	European Central Bank	0
4	Bank	0
6	International Monetary Fund	0
1	Artificial intelligence	1
2	Machine Learning	1
8	Swimming	2
0	Data Science	3
7	Basketball	4
9	Tennis	4
5	Financial technology	5

FIGURE 2.2 Result of clustering

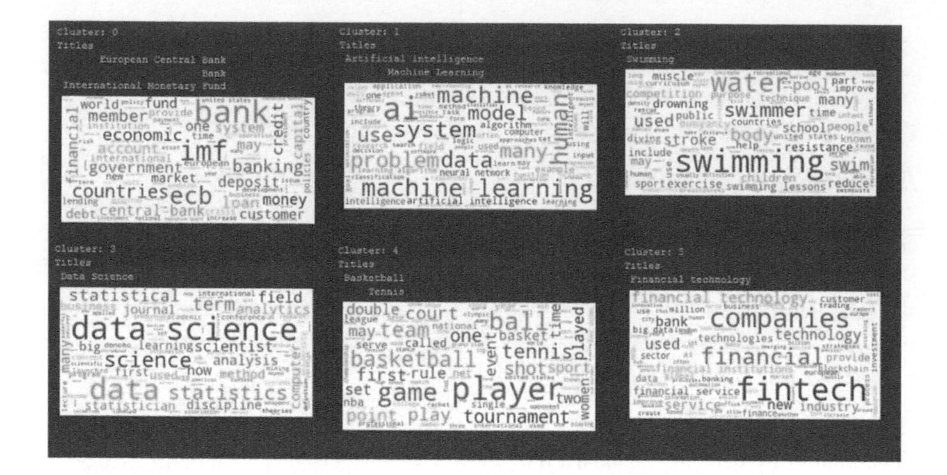

FIGURE 2.3 Word clouds formed as clusters

```
titles=wiki_cl[wiki_cl.cluster==k]['title']
print(titles.to_string(index=False))
plt.figure()
plt.imshow(wordcloud, interpolation="bilinear")
plt.axis("off")
plt.show()
```

Word clouds formed as clusters

It is represented in Figure 2.3.

Cluster 0 consists of articles on European Central Bank, Bank, and International Monetary Fund.
Cluster 1 consists of articles on artificial intelligence and machine learning.
Cluster 2 has an article on swimming
Cluster 3 has an article on data science
Cluster 4 has articles on basketball and tennis
Cluster 5 has an article on financial technology

It might seem odd that swimming is not in the same cluster as basketball and tennis. Or that AI and machine learning are not in the same group with data science. That is because we choose to create six clusters. But by looking at the word clouds, we can see that articles on basketball and tennis have words like game, player, team, and ball in common while the article on swimming does not.

2.2.4 Deep Learning Approach for Text Summarization

Deep learning is a field which attracted many researchers to solve a variety of problems. It is a branch of machine learning which is completely based on artificial neural networks, as neural network is going to mimic the human brain, so deep learning is also a kind of mimic of the human brain. This works well for large datasets. There have been many researchers using deep learning models for both extractive and abstractive text summarization approaches. Extractive summarization methods in particular could be classified into three categories based on deep learning techniques, which are Restricted Boltzmann Machine (RBM), Variation Auto-Encoder, and Recurrent Neural Network. The datasets used for the evaluation are Daily Mail and DUC2002. Some prominent methods are discussed in this section.

Ensemble Noisy Auto-Encoder method was given by Azar et al. (2017). An auto-encoder (AE) is a feed-forward network with the function of learning to reconstruct the input x. The AE is indeed trained to encode the input x using a set of recognition weights into a concept space $C(x)$. A deep AE transforms the original input to a more discriminative representation in coding layer (discriminative layer or bottleneck) over a hierarchical feature detection in which each level corresponds to a different level of abstraction.

The model has two training stages: pre-training and fine-tune. Pre-training provides an appropriate starting point for the fine-tune phase. In pre-training phase, restricted Boltzmann machine (RBM) is used as a model with two layers: one layer visible and the other one hidden. As shown in Figure 2.4, RBM is a generative model.

These RBMs are stacked on top of the other for forming a network. This is shown in Figure 2.5.

After extracting the weights of stacked RBMs from the greedy layer-wise unsupervised algorithm (i.e. pre-training), the weights are used as an initialization point of AE with a corresponding configuration.

The texts are represented using BoW model with TF-IDF being used for word representation. This generates a sparse matrix. A less sparse representation is produced using a local vocabulary to construct a locally normalized term-frequency (tf) representation. In this representation, the vocabulary for each document is separately constructed from the most frequent terms occurring in that document. The vocabulary size is the same for all the documents. This local representation logically has to be less sparse compared to TF-IDF in which the vocabulary is built from all the documents because the input dimensions correspond only to terms occurring in the current document.

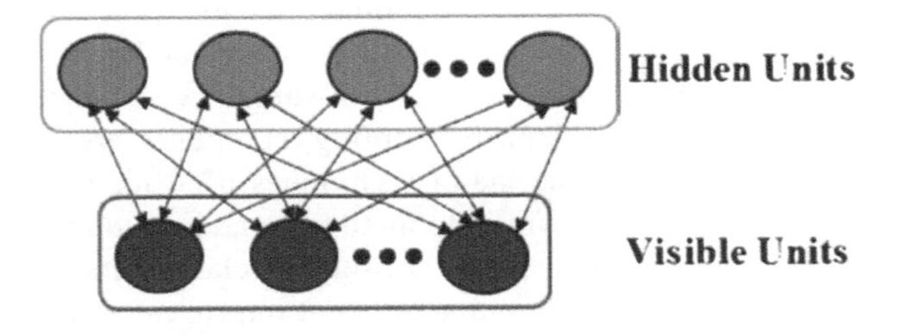

FIGURE 2.4 RBM model with two layers

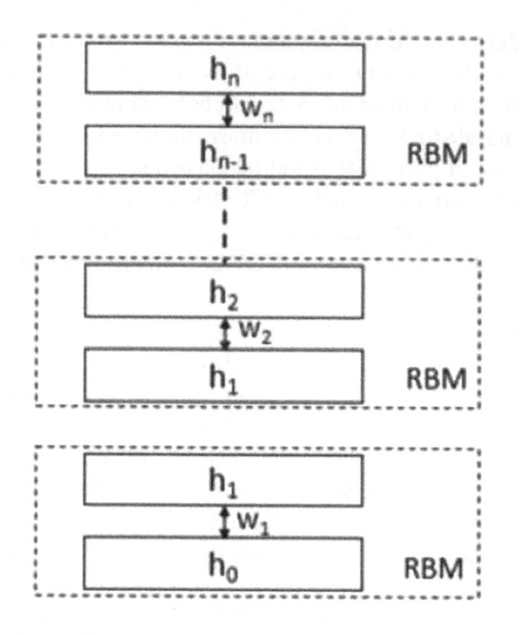

FIGURE 2.5 Stacks of RBMs

The model ranks the sentences based on the most relevant and salient characteristics of the documents that are semantically similar to either the subject of an email or previously selected key phrases of a text; thus, this ranking process is an example of query-oriented summarization. The email subject or key phrase is the query text. The relevance score is computed using cosine similarity between query and input documents. After ranking the sentences of a document based on the cosine distance to the given query, a number of them must be selected to be included in the summary. A straightforward selection strategy is just to select the top-ranked sentences to be included in the summary. The system was able to produce better results.

Subsequent work published on deep learning–based summarization focuses on using classification for summarization. The system extracts the important sentences from a given text. Then, deep learning–based classifier is used to classify the given text as either summary sentences or non-summary sentences. Since the number of summary sentences is very less compared to non-summary sentences, the classifier tends to be biased towards the majority class. Here the majority class is non-summary sentences. To overcome the class imbalance problem, resampling techniques are used prior to the classification task. The three major techniques used for resampling are under-sampling methods, over-sampling methods, and hybrid methods. Deep learning–based multi-layer perceptron is used for classification. Evaluation is performed using CNN dataset, and significant improvement is found compared to baseline methods.

Large-Scale Summarization Methods

Large-scale datasets for summarization suffer from a diversity of datasets and reliability of the summary generated. Some datasets and evaluation methods are elaborated. Mahnaz et al. created a wikiHow dataset, which contains more than 230,000 article and summary pairs extracted and constructed from an online knowledge base written by different human authors. The articles span a wide range of topics and therefore represent high diversity styles. This new large-scale summarization dataset consists of diverse articles from wikiHow knowledge base. This can be used for the evaluation of summarization systems.

Data Extraction and Dataset Construction

A web crawler based on Python was used to get the data from the wikiHow website. The articles classified into 20 different categories cover a wide range of topics.

Data is prepared for the summarization task by considering each method (if any) described in the article as a separate article. To generate the reference summaries, bold lines representing the summary of the steps are extracted and concatenated. The remaining parts of the steps are also concatenated to form the source article. After this step, 230,843 articles and reference summaries are generated. There are some articles with only the bold lines, i.e. there is no more explanation for the steps, so they cannot be used for the summarization task. To filter out these articles, a size threshold is used, so that pairs with summaries longer than the article size will be removed. The final dataset consists of 204,004 articles and their summaries. The statistics of the dataset is shown in Table 2.5.

Evaluation Method

Verma and Patel (2017) recommended two methods for summarizing large reviews and feedback datasets: (i) a graph-based sentence scoring method and (ii) a MiniBatchKMeans clustering algorithm.

Graph-Based Sentence Scoring Algorithm

In this method, sentences are represented by vertices, and similarities between sentences are represented by edges between the vertices. It is an unsupervised learning-based approach for extractive text summarization by automatic sentence extraction using graph-based ranking algorithms. In short, a graph-based ranking algorithm is a way of deciding on the importance of a vertex within a graph.

TABLE 2.5
wikiHow Dataset Statistics

Parameters	Number of Articles/Words
Dataset size	230,84230,843
Average article length	579.8
Average summary length	62.1
Vocabulary size	556,46556,461

In this research work, the vertexes represent the review or feedback given by individual, by taking into account global information recursively computed from the entire graph, rather than relying only on local vertex-specific information. Participating review and feedback in summary sentences are well connected to other sentences. The connectivity of the sentences which is represented by vertexes is based on similarity with other sentences. Similarity measures like TF-IDF are used to measure the performance of the system. Graph is represented by G(V, E), where V is the set of sentences and E the similarity between the sentences. A threshold value is decided for similarity between the sentences. Sentence score is calculated based on the rank of sentences which is estimated by their degree. Top *k* sentences are selected for summarizing sentences.

The results achieved by applying graph-based text summarization techniques with large-scale review and feedback data found improvement with previously published results based on sentence scoring using TF as well as TF-IDF.

A MiniBatchKMeans Clustering Algorithm

MiniBatchKMeans is a modified version of k-means algorithm which is more efficient for large-scale web data. Due to the large size and time constraint for cluster, MiniBatchKMeans performs more effectively compared to k-means. Suppose a dataset of 500,000 reviews and feedbacks is considered, the objective is to divide them into 100 clusters. The complexity of the original k-means clustering algorithm is $O(n*K*I*f)$, where n is the number of records, K is the number of clusters, I is the number of iterations, and f is the number of features. In this research work, review or feedback given by an individual is considered a document. Due to the large size, the small-size subsets are selected from original dataset, and then the algorithms for clustering are applied. The algorithm takes small batches (randomly chosen) of the dataset for each iteration. It then assigns a cluster to each data point in the batch, depending on the previous locations of the cluster centroids.

For k-means:

```
km = KMeans(n_clusters=8, init='k-means++', max_iter=100,
  n_init=1,verbose=0)
```

For MiniBatchKMeans:

```
km = MiniBatchKMeans(n_clusters=8, init='kmeans++',n_init=1,init_
  size=1000, batch_size=1000, verbose=0)
```

Cluster and summary analysis were performed using three approaches. Three approaches used for summary analysis are:

a) Sentence score based on word TF
b) Sentence score using word base form TF
c) Sentence score using word base form TF-IDF

MiniBatchKMeans improves the result compared to k-means and achieves good results. The system used Spark GraphX programing environment which was found to be very effective for handling large-scale data. This shows that unsupervised methods suit well to handle large datasets.

Milenova and Sakharov (2019) recently published a patent for semantic large-scale summarization. First, a large-scale classification is used. This is a two-step process: modeling and classification. In modeling, sparse matrix is constructed such that rows are class vectors and columns are attributes. For classification, input is represented as a case vector of weighted attributes of the case. Similarity is computed for the case vector and the sparse matrix. Now similarity measure above a specified threshold is predicted to be the class for the corresponding case. The text summary approach using explicit semantic analysis operates generally as follows: (i) grammatical units (e.g. sentences or words) are extracted from the given text document using any known technique for identifying and extracting such units, (ii) each of the extracted grammatical units and the text document are represented as weighted vectors of knowledge base concepts, (iii) the semantic relatedness between the text document as a whole and each grammatical unit is computed using the weighted vectors, and (iv) one or more of the grammatical units most semantically related to the text document as a whole are selected for inclusion in a text summary of the text document. The concept vectors are formed for each candidate grammatical unit and the concept vector formed for the document. It selects the plurality of the selected grammatical units, from the plurality of candidate grammatical units, for inclusion in a text summary of the document based on one or more similarity measures computed for the plurality of the selected grammatical units.

Cognitive Models for Summarization

The environment proposed by Benoît et al. implements the two subprocesses as follows: Assessing the importance of sentences in the source text is performed by an implementation of model using LSA. Detecting the use of macro rules and providing a diagnosis were implemented from the theoretical model we presented previously. The interaction with the student is as follows. First, the student is provided with a source text. After reading it, the student writes out a summary of the text in another window. Second, at any time the student can get an assessment of the summary. This feedback may either highlight sentences depending on whether they are adequate or not, or deliver diagnostic messages about the macro rules the student applied.

The main asset of the learning environment is its cognitive foundations. The summarizing process engages numerous complex cognitive skills that have to be assessed in order to assess a summary. A two-step process used was assessing the importance of sentences and detecting the student use of macro rules. Although the first subprocess has been subject to empirical validation, the second remains to be confronted with human data. We thus plan to ask teachers to detect and diagnose macro rule application after reading the student's summaries. This comparison against human data could help tackle a major issue: LSA-based models often rely on similarity thresholds to decide between two alternatives (sentences are coherent or not, words are semantically related or not, etc.). However, it is quite hard to set the value of those thresholds. They are often arbitrarily determined, and we plan instead to perform a fine-tuning by a comparison to human data. The cognitive skills involved in the summarization process probably depend on the nature of the text read.

A novel text saliency model proposed by Sood et al. (2020) uses cognitive reading model for text. This can be further used and applied to text summarization process. The model contains two stages: First, a novel hybrid text saliency model was introduced that, for the first time, integrates a cognitive reading model with a data-driven approach to address the scarcity of human gaze data on text. Second, they proposed a novel joint modeling approach that allows the text summarization (TSM) to be flexibly adapted to different NLP tasks without the need for task-specific ground truth human gaze data.

Summary

Several machine learning methods had been employed for improving the performance of extractive summarization systems. Although the success of these approaches had been evident from the literature, there are a few limitations. Applications of these methods for large-scale data have limitations because of scaling and computation complexity. These algorithms need to be updated for scaling large datasets and to improve the quality of results in the task of summarization systems.

Exercises

1. This exercise explores the quality of the n-gram model of language. Find or create a monolingual corpus of 100,000 words or more. Segment it into words, and compute the frequency of each word. How many distinct words are there? Also count the frequencies of bigrams (two consecutive words) and trigrams (three consecutive words).

   Now use those frequencies to generate language: from the unigram, bigram, and trigram models, in turn, generate a 100-word text by making random choices according to the frequency counts.

   Compare the three generated texts with the actual language. Finally, calculate the perplexity of each model.
2. Consider the following toy corpus:
   the cat cut the hat
   - How many different bigrams of characters (including whitespace) do you have in that corpus?
   - How many occurrences do you have in total? (i.e. including repetitions)
   - Considering only lowercase alphabets and whitespace, how many bigrams are possible?
3. Identify the tokens of the given text document

   Text = 'Last week, the University of Cambridge shared its own research that shows if everyone wears a mask outside home, dreaded "second wave" of the pandemic can be avoided.'

4. Tokenize the given text with stop words ('is,' 'the,' 'was') as delimiters. Tokenizing this way identifies meaningful phrases for topic modeling.
   Text = 'Walter was feeling anxious. He was diagnosed today. He probably is the best person I know.'
5. Express the statement in the form of UNL hypergraph: John is reading a novel.
6. Express the statement as UNL expression: He was director of the Academy.
7. Suppose we have height, weight, and T-shirt size of some customers, and we need to predict the T-shirt size of a new customer – new customer named 'Monica' has height 161 cm and weight 61 kg – given only height and weight information. Data including height, weight, and T-shirt size information is shown in Table 2.6.
   Use KNN classification and predict the target for the new unknown data.
8. The whole process of k-means clustering as per the data mentioned in Table 2.7 can be divided into the following stages:
   Preprocessing of the dataset
   Feature extraction with TF-IDF
   Running k-means and cluster analysis
   Explain the steps in detail.

TABLE 2.6
T-Shirt Size Based on Height and Weight

Height (in cm)	Weight (in kg)	T-Shirt Size
158	58	M
158	59	M
158	63	M
160	59	M
160	60	M
163	60	M
163	61	M
160	64	L
163	64	L
165	61	L
165	62	L
165	65	L
168	62	L
168	63	L
168	66	L
170	63	L
170	64	L
170	68	L

TABLE 2.7
K-means Clustering

	Country	Alpha-2	Alpha-3	Continent	Anthem
0	Albania	AL	ALB	Europe	Around our flag we stand united, With one wish…
1	Armenia	AM	ARM	Europe	Our fatherland, free, independent, That has fol…
2	Austria	AT	AUT	Europe	Land of mountains, land by the river, Land of …
3	Azerbaijan	AZ	AZE	Europe	Azerbaijan, Azerbaijan! The glorious Fatherlan…
4	Belarus	BY	BLR	Europe	We, Belarusians, are peaceful people, Wholehea…
5	Belgium	BE	BEL	Europe	O dear Belgium, O holy land of the fathers …

REFERENCES

Gupta, V. K., and Siddiqui, T. J. 2012, December. "Multi-document summarization using sentence clustering." In *2012 4th International Conference on Intelligent Human Computer Interaction (IHCI)*, Kharagpur, India: IEEE (pp. 1–5).

Jo, D. T. 2019. "String vector based K nearest neighbor for news article summarization." In *Proceedings on the International Conference on Artificial Intelligence (ICAI)*. Athens: The Steering Committee of The World Congress in Computer Science, Computer Engineering and Applied Computing (WorldComp). (pp. 146–149).

Kogilavani, A., and Balasubramani, P. 2010. "Clustering and feature specific sentence extraction based summarization of multi-documents." *International Journal of Computer Science & Information Technology (IJCSIT)* 2(4): 99–111.

Milenova, B., and Sakharov, A. 2019. "Approaches for large-scale classification and semantic text summarization." *U.S. Patent Application* 15/825,930, filed May 30.

Neto, J.L., Freitas, A.A., and Kaestner, C.A.A. 2002. "Automatic text summarization using a machine learning approach." In Bittencourt, G., and Ramalho, G.L. (eds), *Advances in Artificial Intelligence*. Berlin, Heidelberg: Springer. SBIA 2002. Lecture Notes in Computer Science, vol 2507. https://doi.org/10.1007/3-540-36127-8_20.

Prathima, M., and Divakar, H. R. 2018. "Automatic extractive text summarization using K-means clustering." *International Journal of Computer Sciences and Engineering* 6: 782–787. doi: 10.26438/ijcse/v6i6.782787.

Sarkar, K. 2009. "Sentence clustering-based summarization of multiple text documents." *TECHNIA–International Journal of Computing Science and Communication Technologies* 2(1): 325–335.

Shah, C., and Jivani, A. 2019. "An automatic text summarization on naive Bayes classifier using latent semantic analysis." In Shukla, R., Agrawal, J., Sharma, S., and Singh Tomer, G. (eds), *Data, Engineering and Applications*. Singapore: Springer.

Shlens, J. 2005. *A Tutorial on Principal Component Analysis*. San Diego: Systems Neurobiology Laboratory University of California.

Sood, E., Tannert, S., Müller, P., and Bulling, A. 2020. "Improving natural language processing tasks with human gaze-guided neural attention." *Advances in Neural Information Processing Systems* 33: 6327–6341.

Verma, J. P., and Patel, A. 2017. "Evaluation of unsupervised learning based extractive text summarization technique for large scale review and feedback data". *Indian Journal of Science and Technology* 10(17): 1–6.

Yousefi-Azar, M., and Hamey, L. 2017. "Text summarization using unsupervised deep learning." *Expert Systems with Applications* 68: 93–105.

SAMPLE CODE

```
import torch
from torch import nn
from torch.nn import init
from transformers import BertModel, RobertaModel class MatchSum(nn.Module):
def __init__(self, candidate_num, encoder, hidden_size=768):
super(MatchSum, self).__init__()
         self.hidden_size = hidden_size
      self.candidate_num = candidate_num
if encoder == 'bert':
         self.encoder = BertModel.from_pretrained('bert-base-uncased')
else:
         self.encoder = RobertaModel.from_pretrained('roberta-base')
def forward(self, text_id, candidate_id, summary_id):
      batch_size = text_id.size(0)
             pad_id - 0     # for BERT
if text_id[0][0] == 0:
         pad_id = 1 # for RoBERTa
      # get document embedding
      input_mask = ~(text_id == pad_id)
out = self.encoder(text_id, attention_mask=input_mask)[0] # last layer
      doc_emb = out[:, 0, :]
assert doc_emb.size() == (batch_size, self.hidden_size) # [batch_size,
                          hidden_size]
             # get summary embedding
      input_mask = ~(summary_id == pad_id)
out = self.encoder(summary_id, attention_mask=input_mask)[0] # last layer
      summary_emb = out[:, 0, :]
assert summary_emb.size() == (batch_size, self.hidden_size) # [batch_size,
                              hidden_size]
      # get summary score
      summary_score = torch.cosine_similarity(summary_emb, doc_emb, dim=-1)
      # get candidate embedding
      candidate_num = candidate_id.size(1)
      candidate_id = candidate_id.view(-1, candidate_id.size(-1))
      input_mask = ~(candidate_id == pad_id)
out = self.encoder(candidate_id, attention_mask=input_mask)[0]
      candidate_emb = out[:, 0, :].view(batch_size, candidate_num, self.
hidden_size) # [batch_size, candidate_num, hidden_size]
assert candidate_emb.size() == (batch_size, candidate_num, self.hidden_size)
             # get candidate score
      doc_emb = doc_emb.unsqueeze(1).expand_as(candidate_emb)
score = torch.cosine_similarity(candidate_emb, doc_emb, dim=-1) # [batch_
size, candidate_num]
assert score.size() == (batch_size, candidate_num)
return {'score': score, 'summary_score': summary_score}
```

3 Sentiment Analysis Approach to Text Summarization

3.1 INTRODUCTION

Sentiment analysis has been widely used for making key decisions in business. Sentiments for summarization have gained importance since they capture opinions about the entities which are of great interest. Sentiment summarization can be viewed as a subfield of text summarization. The goal of a sentiment summarization system is to generate a summary S, representative of the average opinion about the important aspects relevant to an entity. This can be managed to solve using either a machine learning approach or a natural language processing approach, since it learns sentiments from text using natural language features. This performs fine-grained analysis at the aspect level or feature level and is known as feature-based opinion mining and summarization.

3.2 SENTIMENT ANALYSIS: OVERVIEW

Sentiment analysis is a natural language process, which intends to detect emotions or sentiments from the text. In order to perform sentiment extraction, as a first step the sentences in the natural language have to be classified as subjective or objective. Subjective sentences are sentences that contain real opinions or feelings of the authors or writers. In contrast, objective sentences are the ones which states only the facts on various entities. An entity can be a product, person, event, organization, or topic.

The extracted subjective sentences are subjected to classification as positive, negative, or neutral. This reveals the polarity of classification. After classification, the scores corresponding to positive, negative, or neutral are computed to identify the overall opinion/sentiment of the sentence. These scores are used for summarizing sentences based on an entity with sentiments.

3.2.1 Sentiment Extraction and Summarization

Sentiment extraction plays a key role in sentiment summarization. Capturing real sentiments based on the user's perspective is essential to generate a meaningful summary. This becomes more essential when this can be applied to specific entities of interest. The various stages of the process are shown in Figure 3.1 and explained in detail.

DOI: 10.1201/9781003371199-3

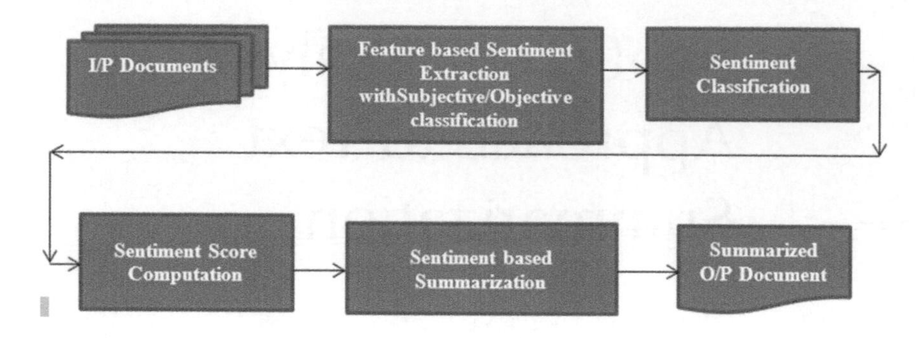

FIGURE 3.1 Stages of feature-based sentiment summarization

3.2.1.1 Sentiment Extraction from Text

The role of sentiment extraction is significant in summarization. Subjective sentences which carry user's beliefs, sentiments, and feelings are more usable for summarization task. This reveals information about significant entities involved for summarization. The identification of subjective sentences can be performed as a classification task when a whole document or entire set of documents are considered. Subjectivity classification can be done using features like term frequency, position modifier, negation words, etc., as a machine learning task. Along with subjectivity classification, another major task in this stage is feature/aspect identification based on part-of-speech (POS) tagging. In this stage POS tags are annotated with the text for identifying sentences with nouns and adjectives. These sentences are used in the classification process.

3.2.1.2 Classification

The sentiments extracted are classified as positive, negative, or neutral. This classification is based on the knowledge words or dictionary formed with standard keywords used for expressing opinions. This knowledge base is also known as the sentiment lexicon or dictionary. Some example terms for positive words are excellent, beautiful, good, etc. Similarly negative words could be bad, horrible, ugly, etc. After classification, these sentiments are aggregated over some entities to determine the overall sentiment. Then scores have to be computed for the sentences over entities for identifying significant sentences to be included in the summary.

3.2.1.3 Score Computation

Subjective sentences with sentiments are scored based on the degree of opinions in the sentence. The final sentiment for an entity is computed from all the sentences which represent that specific entity. Suppose 'camera' is an entity considered in the sentences shown:

- The battery life of this camera is very good.
- Camera is a good device for capturing photographs.

Sentiment scores from the two sentences are aggregated for obtaining the overall score of the entity 'camera.'

3.2.1.4 Summary Generation

The opinions are summarized, and short summary about the entity is created. The entire process is illustrated with an example in the next section.

3.2.2 SENTIMENT SUMMARIZATION: AN ILLUSTRATION

This section demonstrates a practical example of sentiment summarization along with some formal definitions. Consider the review sentence shown in the example below:

> The iPhone 6 feels good to hold, beautifully solid, with a smooth, metal back and glass front. But it has an aura of fragility – maybe it's the extra-slim look, or the massive pane of curved glass on the front. I instantly wanted to slip it into a case just to be safe. Early reports of bending iPhones may havc been overblown. Consumer Reports' testing found that the new iPhones can withstand less pressure than the iPhone 5, LG G3 and Samsung Galaxy Note 3, but are about as equally durable as the HTC One M8. In other words, it's not exactly delicate, but, like any other phone, it can be damaged – so handle with care, and get a case for it.

The steps described in the previous section for sentiment extraction–based summarization are elaborated with the running example, using the review sentence shown above. These steps are presented below:

Step 1: Input preprocessing and classification of sentences as subjective or objective.

Sample Sentiment Lexicon: Any dictionary which holds sentiment words or opinion words.

O/p: Subjective Sentences: The **iPhone 6** feels good to hold, beautifully solid, with a **smooth, metal back and glass front**. But it has an aura of fragility – maybe it's the extra-slim look, or the **massive pane of curved glass** on the front. Consumer Reports' testing found that the new iPhones can withstand less pressure than the iPhone 5, LG G3 and Samsung Galaxy Note 3, but are about as equally durable as the HTC One M8. In other words, it's not exactly delicate, but, like any other phone, it can be damaged – so handle with care, and get a **case** for it.

Features Identified: iPhone 6, back, and front glass, case/cover.

Objective Sentences: I instantly wanted to slip it into a case just to be safe. Early reports of bending iPhones may have been overblown.

The subjective sentences are used as input for sentiment classification.

Step 2: Classification of sentiments:

> The iPhone 6 feels **good to hold, beautifully solid**, with a smooth, metal back and glass front. But it has an aura of fragility – maybe it's the **extra-slim look**, or the

massive pane of curved glass on the front. Consumer Reports' testing found that the new iPhones can withstand **less pressure** than the iPhone 5, LG G3 and Samsung Galaxy Note 3, but are about as equally durable as the HTC One M8. In other words, it's **not exactly delicate**, but, like any other phone, it can be **damaged** – so handle with care, and get a case for it

Sentiment classification is based on the dictionary of positive or negative keywords. This depends on the quality of the dictionary. The subjective sentences are compared with the sentiment lexicon for further classification of sentiments. Words extracted from the above sentences with their dictionary are given below:

Positive keywords: good, extra, delicate, beautifully

Phone: beautifully solid, extra-slim look

Negative keywords: less, damaged

Phone: less pressure, damaged, not exactly delicate

If any positive or negative keyword is preceded by negation words like nothing, not in the dictionary, then it will be categorized as opposite opinion class as appears in the dictionary. In the example above, **not exactly delicate** will be classified as **negative opinion**, since negation word is preceded with positive keyword in the lexicon.

Step 3: Sentiment score computation

Scores are computed for the classified opinions as positive scores or negative scores. SentiWordNet (SWN) is a popular lexicon used for computing sentiment scores of sentences. SentiWordNet is derived from WordNet by assigning positive, negative, and objective scores to all WordNet synsets. Each sentiment score is a real value in the interval [0, 1] which signifies how positive, negative, and neutral each term contained in the synset. Recent version (SentiWordNet 3.0) is enhanced to improve the sentiment score by performing a random walk on WordNet 3.0. Table 3.1 shows the ten top-ranked positive and negative synsets in SentiWordNet 3.0.

The consolidation of scores for all the features is given with a textual summary for the significant features identified.

Step 4: Sentiment-based summarization

Text-based summaries are generated for features along with sentiment scores. This gives the final summary of the significant features extracted. There are two major categories of summaries created based on sentiments. One is text based, and the other is visual summaries of the features. The summarized output is achieved by adding sentiment scores for the features from all the input sentences.

Summarized Output

Sample summary output is shown for iPhone, and its two features like battery and case are given below: Summary is achieved using aggregation of scores on the features battery and case from the entire review input sentences. General indicates the summary scores from the sentences which gives about the overall opinions about iPhone.

TABLE 3.1
Top Ten Positive and Negative Synsets from WordNet

Rank	Positive	Negative
1	good#n#2, goodness#n#2	abject#a#2
2	better off#a#1	deplorable#a#1 distressing#a#2 lamentable#a#1 pitiful#a#2, sad#a#3 sorry#a#2
3	divine#a#6, elysian#a#2 inspired#a#1	bad#a#10, unfit#a#3 unsound#a#5
4	good enough#a#1	scrimy#a#1
5	solid#a#1	cheapjack#a#1 shoddy#a#1 tawdry#a#2

iPhone:

+ **General (136)** – Positive sign indicates 136 reviews are positive sentences regarding iPhone from the entire input review sentences.
– **Battery (276)** – Negative sign indicates 276 reviews are negative regarding iPhone's battery from the entire input review sentences.

The battery life is long (180) – The score 180 indicates the cumulative score for battery life.

The battery was not available anywhere (10) – The score 10 indicates the cumulative score for availability of battery. Since it contains negation word, this will be counted as a negative score.

+ **Case (37)** – This indicates a positive score for case/cover for the iPhone.

As mentioned above, summaries can be generated for all significant entities or features, which provide some reasonable insights about a product or any other matter of interest.

3.2.3 Methodologies for Sentiment Summarization

Feature-based sentiment summarization or aspect-based sentiment summarization can be accomplished using rule-based or machine learning approaches. The summaries generated with these methods can be broadly classified as textual, visual, and statistical summary. This section analyzes the recent techniques based on the literature on aspect-based opinion summarization.

Opinion summarization field became more prevalent from the research explored by Hu and Liu (2004) on the summarization of product reviews. This had been considered a base method for summarizing opinions. The major steps performed are as follows:

Extraction of Frequent Features

Association rule mining using a priori algorithm is used for extracting the frequent features. An itemset is simply a set of words or a phrase that occurs together in some sentences. The main reason for using association mining is the following observation. It is common that a customer review contains many points that are not directly related to product features. Different customers usually have different stories. However, when they comment on product features, the words that they use converge. Thus, using association mining to find frequent itemsets is appropriate because those frequent itemsets are likely to be product features. Those noun/noun phrases that are infrequent are likely to be non-product features. A priori algorithm is run on the transaction set of noun/noun phrases produced from POS tagging. Each resulting frequent itemset is a possible feature. The features resulting from the previous step could contain unlikely features also. Two types of pruning are used to remove those unlikely features.

Feature Pruning

Compactness pruning and redundancy pruning are used to remove unlikely features.

Compactness Pruning

This method checks features that contain at least two words, which can be known as feature phrases, and removes those that are likely to be meaningless. The association mining algorithm does not consider the position of an item (or word) in a sentence. However, in a sentence, words that appear together in a specific order are more likely to be meaningful phrases. Therefore, some of the frequent feature phrases generated by association mining may not be genuine features. Compactness pruning aims to prune those candidate features whose words do not appear together in a specific order.

Redundancy Pruning

Removing redundant features that contain single words is performed in this step. The meaning of redundant features was defined using the concept of p-support (pure support). P-support of feature, ftr, is the number of sentences that ftr appears in as a noun or noun phrase, and these sentences must contain no feature phrase that is a superset of ftr. A minimum p-support value is used to prune those redundant features. If a feature has a p-support lower than the minimum p-support (it is set to 3) and the feature is a subset of another feature phrase (which suggests that the feature alone may not be interesting), it is pruned. For example, life by itself is not a useful feature while battery life is a meaningful feature phrases.

Extraction of Opinion and Opinion Orientation

Opinions are extracted along with their orientation. If a sentence contains one or more product features and one or more opinion words, then the sentence is called an opinion sentence. Thus all opinion sentences are identified. Orientation of opinion is identified using seed words. The strategy is to use a set of seed adjectives, which is already known with their orientations, and then grow this set by searching in the WordNet. Manually a set of very common 30 adjectives are used as the seed list, e.g. positive adjectives: great, fantastic, nice, cool; and negative adjectives: bad, dull. Then WordNet is resorted to predict the orientations of all the adjectives in the opinion word list. Once an adjective's orientation is predicted, it is added to the seed list. Therefore, the list grows in the process. A procedure is executed as follows:

Procedure for orientation search checks WordNet and the seed list for each target adjective word to predict its orientation. Next it searches synset of the target adjective in WordNet and checks if any synonym has a known orientation. If so, the target orientation is set to the same orientation as the synonym, and the target adjective along with the orientation is inserted into the seed list. Otherwise, the function continues to search the antonym set of the target word in WordNet and checks if any antonym has a known orientation. If so, the target orientation is set to the opposite of the antonym, and the target adjective with its orientation is inserted into the seed list. If neither synonyms nor antonyms of the target word have known orientation, the function just continues the same process for the next adjective since the word's orientation may be found in a later call of the procedure with an updated seed list. For those adjectives that WordNet cannot recognize, they are discarded as they may not be valid words. For those adjectives, orientations cannot be found, and they will also be removed from the opinion word list. Now orientations of opinions have to be predicted for sentences.

Predicting the Orientations of Opinions

The next step is predicting the orientation of an opinion sentence, i.e. positive or negative. The dominant orientation of the opinion words in the sentence is used to determine the orientation of the sentence. That is, if positive/negative opinion prevails, the opinion sentence is regarded as a positive/negative one. In the case where there is the same number of positive and negative opinion words in the sentence, prediction of the orientation is performed by using the average orientation of effective opinions or the orientation of the previous opinion sentence. Finally summary sentences are generated for a specific feature with score, positive comments, and negative sentences.

Summary Generation

This consists of the following two steps:

(i) For each discovered feature, related opinion sentences are put into positive and negative categories according to the orientations of the opinion sentences. A count is computed to show how many reviews give positive/negative opinions about the feature.

(ii) All features are ranked according to the frequency of their appearances in the reviews. Feature phrases appear before single-word features as phrases normally are more interesting to users.

Rule-Based Methodologies

Aspect extraction is the main task in aspect-based opinion summarization. There are two types of aspects identified in this task: implicit and explicit aspects. Rule-based approaches like association rule mining and clustering have been widely adapted for extracting implicit aspects. Dependency rules and lexicons such as WordNet and SentiNet have also been used to identify implicit aspects. The quality of the dictionary or the lexicon plays a key role in extracting the implicit aspects. Recent studies use sequential pattern mining for the extraction of aspects. Rule-based approaches were extensively used for aspect extraction. In rule-based approaches, learning from rule needed extensive training on specific domain which led researchers to explore extraction of aspects in machine learning techniques.

Machine Learning Methods

There are many machine learning methods available for extraction of aspects. Topic modeling such as latent Dirichlet allocation (LDA) and neural network–based approach using convolutional neural networks (CNN) have been extensively used in present-day research.

Beyond these methods, techniques utilizing multi-aspect summarization and abstractive summarization have also been explored. Multiple facets or aspects are considered to be crucial in this field. Mukherjee and Joshi (2013) have shown that a review may have multiple facets with a different opinion about each facet. Traditional sentiment analysis systems ignore the association between the features of the given product domain. Abstractive summarization of aspect-based opinions has also attracted the researchers. The main focus was to generate short summaries which give useful information about features and its opinions. Though there are numerous methods available for aspect-based text summarization with sentiments, only a few researchers have explored the consequences of using sentiments in text summaries.

3.3 IMPLICATIONS OF SENTIMENTS IN TEXT SUMMARIZATION

This section elaborates on the efficiency of using sentiments in text summarization. It has been very crucial from the end user's perspective. Inclusion of sentiments in textual summaries can be very effective when used for analysis. Yadav and Chatterjee (2016) studied the use of sentiments in text summarization. They used a computationally efficient technique using sentiments of keywords in the text for summarization. Sentiment analysis can also be used efficiently for the purpose of text summarization. In this method, sentiment scores and POS tags are identified for a sentence. All the tags identified as adjectives are computed with positive and negative scores associated with that particular term. Based on the scores computed for all sentences, summary sentences are extracted. This is used to compute precision and recall for 10–50% of summary compared to the original input text. This had been found efficient for 50% summarization upon the whole document length.

The research by Dabholkar et al. (2016) uses a degrading extraction approach to create a document summarization framework wherein if one extraction strategy fails, then the model gracefully degrades to another. This system creates virtual paragraphs from the sentiment values. Then it performs aggregation of objective and subjective sentences based on a relative score. The method is elaborated as below.

Sentiment score computation for sentences and documents is performed. If the number of sentences with a sentiment score is more than a threshold, then a virtual paragraph is created using the sentiment terms. A relative score between the original document and the virtual paragraph is computed. Then a subjective/objective categorizer is used for separating the sentences. Aggregation is performed to extract the subjective sentences, objective sentences, and the position of the sentence to create a summary by modifying the virtual paragraph, and this is used as a final summary. Also relative score between this summary and the original document is verified to be maximum as it gives the original meaning of the input document.

Another extractive-based text summarization method is developed by Siddhaling which involves pronoun replacement with proper noun and formation of text summary. Valence Aware Dictionary for sEntiment Reasoning (VADER) (2014) is used for computing information about positive, compound, negative, and neutral scores along with the count of neutral, negative, and positive words in the text. It is a simple rule-based model for general sentiment analysis. VADER sentiment lexicon has been compared to six other well-established sentiment analysis lexicons: Linguistic Inquiry Word Count (LIWC), General Inquirer (GI), Affective Norms for English Words (ANEW), SentiWordNet (SWN), SenticNet (SCN), and Word-Sense Disambiguation (WSD) using WordNet. It outperforms all other dictionaries available. In the work proposed by Kabadjov et al. (2011), how good a sentence is for summarization was evaluated by using the output of a sentiment analyzer. It was tested with different hypotheses on a corpus of blog threads from different domains. The advancements which have initiated the research for text summarization in many global and local languages have been discussed by Dhawale et al. (2020). Many researchers (Kavita et al., 2012; 2010; Mukherjee et al., 2013; Mukherjee and Bhattacharyya 2012; Mukherjee and Joshi 2014; Musto et al., 2019; Rana and Cheah 2018; Urologin 2018) provided works on text summarization.

Cataldo et al.'s approach is to conceive the justification as a summary of the most relevant and distinguishing aspects of the item, automatically obtained by analyzing its reviews. Such a summary is finally presented to the target user as a justification of the received recommendation.

Cognition-Based Sentiment Analysis and Summarization

Cognitive features for improving sentiment analysis were given by Mishra et al. (2017). They combined traditional sentiment features with (a) different textual features used for sarcasm and thwarting detection and (b) cognitive features derived from readers' eye movement behavior. The combined feature set improves the overall accuracy over the traditional feature set–based SA. It is found to be significantly effective for text with complex constructs. In future, this can be further explored by using (a) devising deeper gaze-based features and (b) multi-view classification using independent learning from linguistics and cognitive data. This type of cognitive features can extensively help to improve sentiment-based summarization.

3.4 SUMMARY

Rule-based methods yield good results for sentiment summarization. This is based on its simplicity. Some of the obvious drawbacks of these systems are generation and maintenance of rules for all word combinations, and it is highly domain dependent. These systems are not considering negation words and intense words. In contrast, machine learning techniques are complex, but perform better than rule-based systems. In general, for any sentiment summarization system, it would be better to use a rule-based approach that also leverages machine learning for improving its performance.

Some of the major issues in aspect-based text summarization are:

Identification of implicit aspects based on domain-specific lexicons, length of the summaries for the aspects, representation of summaries – visual or text based – and quality of the summary generated. Combination of sentiments in the summarization improves readability among the customers in their perception. But the summary generated including sentiments is more biased towards opinionated sentences. Some of the significant sentences, which represent key concepts, may get eliminated. This is the major drawback when sentiments are included in the summary.

PRACTICAL EXAMPLES

EXAMPLE 1

Perform sentiment analysis for the following review text.

I **love** this movie! It's **sweet**, but with **satirical** humor. The dialogs are **great** and the adventure scenes are **fun**. It manages to be **romantic** and **whimsical** while laughing at the conventions of the fairy tale genre. I would **recommend** it to just about anyone. I have seen it several times and I'm always **happy** to see it **again**....

Table 3.2 represents the bag of words in this example.

A bag-of-words representation of a document does not only contain specific words but all the unique words in a document and their frequencies of occurrences. A bag is a mathematical set here, so by the definition of a set, the bag does not contain any duplicate words.

This representation in Table 3.3 is also known as corpus. This training set should be easy to interpret.

TABLE 3.2
Table Representing Bag of Words

great	2
love	2
recommend	1
laugh	1
happy	1
...	...

TABLE 3.3
Corpus Representation

Document	w1	w2	w3	w4	...	wn	Sentiment
d1	2	1	3	1		1	Positive
d2	1	5	5	5		1	Negative
d3	3	8	6	8		2	Positive
d4	2	5	1	5		3	Positive
d5	3	0	3	0		0	Negative
d6	0	0	0	0		0	Negative
d7	2	0	0	0		0	Positive
d8	9	2	9	2		2	Negative
...	...	...	...	...	...	...	...

All the rows are independent feature vectors containing information about a specific document (movie reviews), the particular words, and its sentiment. Note that the label sentiment is often denoted as (+, –) or (+ve, –ve). Also, the features w1, w2, w3, w4, …, wn are generated from a bag of words, and it is not necessary that all the documents will contain each of these features/words. These feature vectors are passed to the classifier model for sentiment classification.

Example 2

In this example, a sentiment analysis task from Kaggle competition posted in 2014 and their results are discussed.

‘There’s a thin line between likably old-fashioned and fuddy-duddy, and The Count of Monte Cristo … never quite settles on either side.’

The Rotten Tomatoes movie review dataset is a corpus of movie reviews used for sentiment analysis, originally collected by Pang and Lee [1]. In their work on sentiment treebanks, Socher et al. [2] used Amazon’s Mechanical Turk to create fine-grained labels for all parsed phrases in the corpus. This competition presents a chance to benchmark your sentiment analysis ideas on the Rotten Tomatoes dataset. **You are asked to label phrases on a scale of five values: negative,**

somewhat negative, neutral, somewhat positive, and positive. Obstacles like sentence negation, sarcasm, terseness, language ambiguity, and many others make this task very challenging. (Note: Implementation is done using GraphLab.)

Table 3.4 is the result from GraphLab. Here the sentiments are converted into two category variables (with or without sentiment more than 3).

A sample model created using a logistic classifier is shown below:

Sample Code (Run Using GraphLab)

```
sf_train['word_count'] = graphlab.text_analytics.count_words
(sf_train['Phrase'])
train,test = sf_train.random_split(0.6,seed=0)
len(train)   93517
basic_model = graphlab.logistic_classifier.create(train,tar
get='target',features = ['word_count'])
```

Figure 3.2 represents the execution in Graph Lab.

Example 3

Identify the aspects, sentiment phrase, and polarity as triples from the following sentences:

The camera on my <xyz-brand> phone sucks. Apart from that, I'm happy.
Aspect: camera phone overall
Sentiment phrase: sucks happy
Polarity: - +
Good price. Sharp image. Vivid colors. Static in Audio. Motion lags a bit.

TABLE 3.4
Result from GraphLab

Phrase Id	Sentence Id	Phrase	Sentiment	Target
112	3	sitting	2	0
113	3	through this one	2	0
114	3	through	2	0
115	3	this one	2	0
116	3	One	2	0
117	4	A positively thrilling combination of	3	0
118	4	A positively thrilling combination of	4	1
119	4	A positively thrilling combination of	4	1
120	4	A positively thrilling combination of	3	0
121	4	A positively thrilling combination of	3	0

```
In [84]: basic_model.evaluate(test)

Out[84]: {'accuracy': 0.94056888860464,
          'auc': 0.8288266846827667,
          'confusion_matrix': Columns:
                 target_label    int
                 predicted_label int
                 count   int

          Rows: 4

          Data:
          +--------------+-----------------+-------+
          | target_label | predicted_label | count |
          +--------------+-----------------+-------+
          |      0       |        1        |  1542 |
          |      1       |        0        |  2175 |
          |      1       |        1        |  1456 |
          |      0       |        0        | 57370 |
          +--------------+-----------------+-------+
          [4 rows x 3 columns],
          'f1_score': 0.43928194297782475,
          'log_loss': 0.24097548441276082,
          'precision': 0.485657104736491,
          'recall': 0.4009914624070504,
```

FIGURE 3.2 Execution in GraphLab

Aspect:	price	image	colors	audio	motion
Sentiment phrase:	good		sharp	vivid	static lags a bit
Polarity: +	+	+	—	-	

REFERENCES

https://www.cnet.com/reviews/apple-iphone-6-review/

Dabholkar, S., Patadia, Y., and Dsilva, P. 2016. "Automatic document summarization using sentiment analysis." In *ICIA-16*. doi: 10.1145/2980258.2980362.

Dhawale, A. D., Kulkarni, S. B., and Kumbhakarna, V. 2020. "Survey of progressive era of text summarization for Indian and foreign languages using natural language processing." In Raj, J., Bashar, A., and Ramson, S. (eds), *Innovative Data Communication Technologies and Application. ICIDCA 2019. Lecture Notes on Data Engineering and Communications Technologies*, vol 46. Heidelberg: Springer.

Hu, M., and Liu, B. 2004, August. "Mining and summarizing customer reviews." In *Proceedings of the ACM SIGKDD international Conference on Knowledge Discovery and Data Mining (KDD)* (pp. 168–177). NewYork, US.

Hutto, C.J. and Gilbert, E.E. 2014. "VADER: A parsimonious rule-based model for sentiment analysis of social media text". *Eighth International Conference on Weblogs and Social Media (ICWSM-14).* Ann Arbor, MI. June 2014.

Kabadjov, M., Balahur, A., and Boldrini, E. 2011. "Sentiment intensity: Is it a good summary indicator?" In Vetulani, Z. (ed.), *Human Language Technology. Challenges for Computer Science and Linguistics. LTC 2009. Lecture Notes in Computer Science*, vol. 6562. Berlin, Heidelberg: Springer.

Kavita, G., ChengXiang, Z., and Evelyne, V. 2012. "Micropinion generation: An unsupervised approach to generating ultra-concise summaries of opinions." In *Proceedings of the 21st International Conference on World Wide Web* (pp. 869–878). Lyon, France.

Kavita, G., ChengXiang, Z., and Jiawei, H. 2010. "Opinosis: A graph based approach to abstractive summarization of highly redundant opinions." In *Proceedings of the 23rd International Conference on Computational Linguistics* (pp. 340–348) Beijing, China: Tsinghua University Press.

Mishra, A., Kanojia, D., Nagar, S., Dey, K., and Bhattacharyya, P. 2017. "Leveraging cognitive features for sentiment analysis." *arXiv preprint arXiv:1701.05581.*

Mukherjee, S., Basu, G., and Joshi, S. 2013, May. "Incorporating author preference in sentiment rating prediction of reviews." In *Proceedings of 22nd International Conference on WWW Companion* (pp. 47–48) New York, NY, United States: Association for Computing Machinery.

Mukherjee, S., and Bhattacharyya, P. 2012. "Feature-specific sentiment analysis for product reviews." In *Computational Linguistics and Intelligent Text Processing* (pp. 475–487). Berlin, Heidelberg: Springer-Verlag.

Mukherjee, S., and Joshi, S. 2013, October. "Sentiment aggregation using concept net ontology." In *Proceedings of 7th International Conference on Natural Language Processing* (pp. 570–578). Nagoya, Japan.

Mukherjee, S., and Joshi, S. 2014. "Author-specific sentiment aggregation for polarity prediction of reviews." In *Proceedings of Ninth International Conference on Language Resources and Evaluation* (pp. 3092–3099).

Musto, C., Rossiello, G., de Gemmis, M., Lops, P., and Semeraro, G. 2019. "Combining text summarization and aspect-based sentiment analysis of users' reviews to justify recommendations." In *Proceedings of the 13th ACM Conference on Recommender Systems (RecSys'19). Association for Computing Machinery*, New York (pp. 383–387). doi: 10.1145/3298689.3347024.

Rana, T., and Cheah, Y.-N. 2018. "Sequential patterns rule-based approach for opinion target extraction from customer reviews." *Journal of Information Science,* 45(5): 643–655.

Urologin, S. Nayak, J., and Satish, L. 2017. "A method to generate text summary by accounting pronoun frequency for keywords weightage computation." *In IEEE, Scopus Indexed, The International Conference on Engineering and Technology*, Turkey August (pp 21–23).

Urologin, S. 2018. "Sentiment analysis, visualization and classification of summarized news articles: A novel approach." *International Journal of Advanced Computer Science and Applications (IJACSA)*, 9(8): 616–625.

Yadav, N., and Chatterjee, N. 2016. "Text summarization using sentiment analysis for DUC data." In *2016 International Conference on Information Technology (ICIT)* (pp. 229–234). Bhubaneswar. doi: 10.1109/ICIT.2016.054.

SAMPLE CODE

//OPINION SCORE CALCULATION

```
import java.io.BufferedReader;
import java.io.BufferedWriter;
import java.io.File;
import java.io.FileNotFoundException;
import java.io.FileReader;
import java.io.FileWriter;
import java.util.ArrayList;
import java.util.HashMap;
import java.util.Iterator;
```

```
import java.util.Set;
import java.util.Vector;
public class opinionExtract {
private String pathToSWN = "sentiment/SentiWordNet.txt"; static
        ArrayList<Double> lst=new ArrayList<Double>(); static
        ArrayList<String> wrds=new ArrayList<String>(); private
        HashMap<String, Double> _dict;
public static HashMap<String,String>pol=new HashMap(); public
        opinionExtract() {
dict = new HashMap<String, Double>();
HashMap<String, Vector<Double>> _temp = new HashMap<String,
Vector<Double>>(); try {
        BufferedReader csv = new BufferedReader(new FileReader(pathToSWN));
String line = "";
while ((line = csv.readLine()) != null) { String[]
        data = line.split("\t");
Double score = Double.parseDouble(data[2]) - Double.parseDouble(data[3]);
String[] words =
        data[4].split(" ");
for (String w : words) { String[]
        w_n = w.split("#");
w_n[0] += "#" + data[0];
int index = Integer.parseInt(w_n[1]) - 1; if
        (_temp.containsKey(w_n[0])) {
        Vector<Double> v = _temp.get(w_n[0]); if
        (index > v.size()) {
for (int i = v.size(); i < index; i++) {
v.add(0.0); } }
        v.add(index,
        score);
temp.put(w_n[0], v); } else { Vector<Double> v = new
        Vector<Double>(); for (int i = 0; i < index; i++)
        {
        v.add(0.0); } v.add(index,score);

temp.put(w_n[0], v); } }} Set<String> temp =
        _temp.keySet();
for (Iterator<String> iterator = temp.iterator(); iterator.hasNext();) {
String word =
        (String) iterator.next();
Vector<Double> v = _temp.get(word); double
        score = 0.0; double sum = 0.0; for (int i
        = 0; i < v.size(); i++) {
score += ((double) 1 / (double) (i + 1)) * v.get(i); } for (int i = 1; i
        <= v.size(); i++) {
sum += (double) 1 / (double) i; } score /=
        sum; String sent = "";
if (score >= 0.75) { sent = "strong_positive";

} else if (score > 0.25 && score <= 0.5) { sent =
        "positive";
} else if (score > 0 && score >= 0.25) { sent =
        "weak_positive";
} else if (score < 0 && score >= -0.25) {
sent = "weak_negative";
} else if (score < -0.25 && score >= -0.5) { sent =
        "negative";
} else if (score <= -0.75) { sent =
        "strong_negative"; }
dict.put(word, score); pol.put(word, sent);}
} catch (Exception e) { e.printStackTrace(); }} public
        Double extract(String word) {
Double total = new Double(0);
```

```
if (_dict.get(word + "#n") != null) { total =
        _dict.get(word + "#n") + total; } if
        (_dict.get(word + "#a") != null) { total =
        _dict.get(word + "#a") + total; } if
        (_dict.get(word + "#r") != null) { total =
        _dict.get(word + "#r") + total; } if
        (_dict.get(word + "#v") != null) { total =
        _dict.get(word + "#v") + total;}
return total; } public static void main(String[] args) throws Exception {
DBConnection db = new
        DBConnection();
void opextract(String fn1,String fn2) throws Exception { opinionExtract
        test = new opinionExtract();
File f2=new File("opstrength");f2.mkdir();
File f3=new File("opstrength/"+fn1);f3.mkdir();
BufferedWriter bw=new BufferedWriter(new FileWriter("opstrength/"+fn1+"/"+fn2));
BufferedReader br = new BufferedReader(new
        FileReader("sentiwordsExtr/"+fn1+"/"+fn2));
String sentence; while ((sentence = br.readLine()) != null) { String[] words
        = sentence.split("\n");
double totalScore = 0; wrds.add(sentence); for (String
        word : words) {
word = word.replaceAll("([^a-zA-Z\\s])", ""); if
        (test.extract(word) == null) { continue; }
        totalScore += test.extract(word);}
lst.add(totalScore); bw.write(sentence+"\t"+totalScore);
int x = db.executeUpdate("insert into opstrength values('" + fn2 + "','" +
sentence + "'," + totalScore +")");
        bw.newLine();
System.out.println(sentence+"\t"+totalScore);
        System.out.println(totalScore); } bw.close(); }}
```

4 Text Summarization Using Parallel Processing Approach

4.1 INTRODUCTION

Parallel processing approaches can be generally categorized as compute-intensive, data-intensive, or both. Data-intensive applications face two major issues: processing exponentially growing data sizes and significantly reducing data analysis cycles with the aim of making timely decisions. While compute-intensive is used to describe application programs that are compute-bound, such applications devote most of their execution time to computational requirements as opposed to input/output and typically require small volumes of data. Finally, data-intensive and compute-intensive applications usually have both features. Parallel applications for text summarization became essential in today's world of increasing data. Parallelization of existing algorithms can be done in two ways: (i) parallelizing computational tasks and (ii) parallelizing for distributed data.

Parallelizing Computational Tasks

Tasks that have no dependencies can be easily parallelized, and iterative computations can also be easily parallelized. These are task dependent and algorithm dependent, which makes it difficult to parallelize.

Parallelizing for Distributed Data

Distributed data requires common distributed storage for parallel execution. Big data paradigm is used commonly with MapReduce frameworks (runs across many machines). MapReduce works best if very small data moves across these machines very few times. This model has been successful for parallel implementation and widely attracted researchers from many fields. Text summarization has also gained importance for parallelization using MapReduce algorithms.

4.2 PARALLEL PROCESSING APPROACHES

Parallel processing approaches widely use parallel algorithms. A parallel algorithm is an algorithm that can execute several instructions simultaneously on different processing devices and then combine all the individual outputs to produce

DOI: 10.1201/9781003371199-4

the final result. The model of a parallel algorithm is developed by considering a strategy for dividing the data and processing method and applying a suitable strategy to reduce interactions. In this chapter, the following parallel algorithm models for summarization are discussed:

Data parallel model
Master slave model

4.2.1 Parallel Algorithms for Text Summarization

Machine learning algorithms can be parallelized depending on platform, communication model, and the size of the data for processing. Some of the algorithms used for text summarization in the literature are: parallel k-means, parallel particle swarm optimization (PSO), and parallel annotation of data with parallel bisection-based k-means. The various methods are discussed in detail.

4.2.2 Parallel Bisection k-Means Method

The multi-document summarization for Arabic text by Maria et al. (2018) uses parallel bisecting k-means algorithm. The phases in the system are detailed.

Phase 1: Document feeding
In this phase, input documents are fed into the system.
Phase 2: Extraction and preprocessing
This phase extracts single or multi-documents from the input data. Then preprocessing is performed.
Phase 3: Parallel bisecting k-means document clustering

This phase implements parallel bisecting k-means document clustering, which is detailed later. Results are generated as a final output of this phase.

Input documents are fed and preprocessed. During preprocessing stage, all possible terms in the input document collection were collected using n-grams (1, 2, 3, and 4 gram words), and frequency of each n-gram occurrence is computed. Arabic WordNet dictionary was used to find the synonyms. The n-gram profile of each text document was compared against the profiles of all documents in corpus in terms of similarity using Manhattan distance.

The bisecting k-means algorithm starts with a single cluster of all documents. It works in the following way:

1) Pick a cluster to be split.
2) Find two sub-clusters using the basic k-means algorithm (bisecting step).

3) Repeat step 2, the bisecting step, for ITER times and take the split which produces the clustering with the highest overall similarity.
4) Repeat steps 1–3 until the desired number of clusters is reached.

The algorithm is shown as steps below:

Input:
C: denotes the number of initial centroids
K: denotes the number of clusters required
MaxIterations: denotes the highest number of k-means iterations at each step
MinDivisibleClusterSize: denotes the lowest number of points (if ≥1.0) or the lowest
Proportion of points (if <1.0) of a divisible cluster (default: 1)
D: denotes a dataset that can be clustered, which contains *n* data points
Output: A set of K clusters
Method:

1. Disseminate *n* data points of top processors in an evenly manner.
2. Determine a cluster Kj to split based on a rule, and send this information to the entire processors.
3. Search for two sub-clusters of Kj by using the k-means algorithm (bisecting steps):
   (a) Determine two data points of Kj to form initial cluster centroids and send them to the pj, processors that contain data members of Kj.
   (b) Every processor performs a calculation for the clustering criterion function of its relevant data points of Kj with two centroids and puts every data point according to its best choice (calculation step).
   (c) Collect the required information entirely in order to update two centroids, and send them to the pj processors, which are involved to participate in the bisecting (update step).
   (d) Repeat steps 3b and 3c until achieving convergence.
4. Repeat steps 2 and 3, and determine the split that obtains the clusters, which satisfy the global function.
5. Repeat steps 2–4 until *k* clusters are acquired.

The process implemented in the steps 3–5 in parallel bisecting k-means algorithm is illustrated as follows:

Extraction of noun and verb key phrases: Extract noun and verb key phrases that have high similarity with user/topic keywords query. Match the frequently occurring noun/verb phrases in all documents in the clusters. Finally, use key phrase features to rank them.

Extraction of Significant Sentences: Extract the most remarkable sentences from each cluster after splitting the cluster into several paragraphs and sentences when using delimiters (e.g. full stop and question mark). Then eliminate the redundancy by using similarity measurements (semantic (25) and syntactic similarity measurements with the help of Arabic WordNet between sentences were used).

Second Clustering Stage: All of the remarkable extracted sentences are fed into another clustering stage using similarity measures between them. Then they are given to summary builder checker, which re-ranks them based on fully coherent, grammatical, and meaningful Arabic sentences.

Summary Builder: The summary builder starts processing by selecting the best scoring sentences from all clusters. Then re-score them according to a modified Arabic language features (which includes extracting all possible events, names, times, etc.). Rank them by identifying the temporal relations between a pair of events in the same sentence.

The summary builder also tests the contents of the sentences by applying the coherent and readability measures to re-ranking them and to selecting the best position for each sentence in the summary, which guarantees that the final summary should not hold non-textual items or punctuation errors (grammaticality).

Phase 8: The final summary is generated from the system. It should not hold excessive information (conquer redundancy) and unclear names and pronouns of people or things without correct reference (reference clarity). Finally, the summary should be in a good structure, and sentence sequence should be coherent. The algorithm is implemented in the environment with parallel implementation using Message Passing Interfaces (MPI) and Python. This algorithm divides workload for multiple central processing units (CPU). Datasets are divided into *n* data points, and they are split evenly based on their size. Four processors run in parallel to process the data based on the algorithm.

There are few approaches explored other than clustering in employing evolutionary approach using parallel algorithms for summarization. Most popular among evolutionary algorithms are genetic algorithms (GA). Meena and Gopalani (2015) gave a new feature set and fitness function for text summarization using GA. Wang and Tang (2012) offered a Chinese automatic summarization method based on Concept-Obtained and Hybrid Parallel Genetic Algorithm. The idea behind this approach is to obtain concepts of words using HowNet as tool and use concepts as features instead of words. A conceptual vector space model is constructed and a Hybrid Parallel Genetic Algorithm is used to form a summary of documents. The automatic evaluation of document summaries with n-gram score shows the effectiveness and feasibility of the system.

In the present-day scenario, the growth of distributed and parallel data processing frameworks paved way for the development of new algorithms for summarization. Most of them try to use MapReduce or Message Passing Interfaces (MPI) from Python. A few significant methods have been discussed in the following section.

4.3 PARALLEL DATA PROCESSING ALGORITHMS FOR LARGE-SCALE SUMMARIZATION

Parallel data processing algorithms are predominantly centered on single-document summarization. Newer researchers in the area of multi-document summarization produce summaries from clusters of similar documents using MapReduce [4]. Some important methods found in the literature are explained along with the concepts in the subsequent sections.

4.3.1 Designing MapReduce Algorithm for Text Summarization

MapReduce (2016) is an open-source data processing framework which generally comes from its simplicity. In addition to preparing the input data, the programmer needs only to implement the mapper, the reducer, and optionally the combiner and the partitioner. All other aspects of execution are handled transparently by the execution framework on clusters ranging from a single node to a few thousand nodes, over datasets ranging from gigabytes to petabytes.

Synchronization is the complex aspect of designing MapReduce algorithms (or for that matter, parallel and distributed algorithms in general). In a single MapReduce job, synchronization opportunity is available only once during the shuffle and sort stage. In this stage intermediate key–value pairs are copied from the mappers to the reducers and grouped by key. Beyond that, mappers and reducers run in isolation without any mechanisms for direct communication.

In designing MapReduce algorithms, the programmer has only little control over various aspects of execution. They are:

- Execution of mapper or reducer node
- Beginning or ending of mapper or reducer
- Type of input key–value pairs processed in a mapper
- Type of intermediate key–value pairs processed in a reducer

The programmer has a number of techniques for controlling execution and managing the flow of data in MapReduce. They are:

1. The ability to construct complex data structures as keys and values to store and communicate partial results.
2. The ability to execute user-specified initialization code at the beginning of a map or reduce task, and the ability to execute user-specified termination code at the end of a map or reduce task.

3. The ability to preserve state in both mappers and reducers across multiple input or intermediate keys.
4. The ability to control the sort order of intermediate keys, and therefore the order in which a reducer will encounter particular keys.
5. The ability to control the partitioning of the key space, and therefore the set of keys that will be encountered by a particular reducer.

The algorithms designed for text summarization depend on clustering techniques. A new parallel k-means clustering approach based on MapReduce framework for aspect-based summary generation is given by Priya et al. (2016). This method incorporates bagging and ensembling techniques for improving efficiency and eliminates noise. The phases involved are feature extraction, classification, and summarization using MapReduce. The features or aspects for which the summary has to be generated are extracted using the LSA technique. Then sentiment scores are computed using naïve Bayes classification algorithm. The sentences with significant feature terms and sentiment scores are aligned in a sequence file to process using MapReduce.

4.3.2 Key Concepts in Mapper

Mapper algorithm includes key logic in the implementation of MapReduce algorithm. In the extractive summarization process, the key task is the extraction of relevant sentences from the input documents. So, in this algorithm, mapper phase is used for the extraction of feature-specific relevant sentences using scores. The algorithm is given below:

Input: Global variable centers, the offsetkey [text values], the samplevalue [review texts], The number of clusters, k

Output: <key’, value’> pair, where the key’ is the index of the closest center point and value’ is a string comprising sample information

1. For each (key in HDFS) do.
2. Compute sentence relevance score with Equation 4.1:

$$s_a(i) = s_t(w,s) + t_r(s) \tag{4.1}$$

where, $s_a(i)$ = score for the sentence for aspect i

$s_t(w, s)$ = average sum of term score values of each distinct term in the sentence

$t_r(s)$ = no of distinct terms occurring in the sentence

3. Based on the score obtained, clusters are formed based on aspect key and their relevant sentences, for ‘k’ clusters.

4. For each cluster change, the centroid with the maximum sentence score is obtained from Equation 4.1.
5. End For.
6. Take each key as the aspect key and value as sentence with aspect relevance score.
7. Output <key, value> pair;
8. End

The mapper algorithm works based on the following steps:

1. The input dataset is stored on Hadoop Distributed File System (HDFS) as a sequence file of <key, value> pairs, each of which represents a record in the dataset.
2. The key is the text value (aspect) from the start point of the file, and the value is a string of the content of this record, which corresponds to the relevant reviews for one particular aspect (key).
3. The dataset is split and globally broadcast to all mappers. Consequently, the distance computations are executed in parallel.
4. For each map task, k-means constructs a global variant center which is an array containing information about the centers of the clusters and their corresponding values.
5. Given the information, a mapper can retrieve the reviews which are related to the aspect and also the relevance score for the sentences retrieved. The output values are then composed of two parts: the text key with the closest center point and the sample information.
6. This output is given as input to the reducer.

Reducer in MapReduce algorithm aggregates the outputs from all the mappers. In the extractive summarization problem, the reducer aggregates the summary sentences based on feature-based sentence score, to generate summary sentences in each cluster. The algorithm and key concepts in reducer are detailed below:

4.3.3 Key Concepts in Reducer

Algorithm for reducer is given below:

Input: key is the aspect key of the cluster; V is the list of sentences with relevance score.

Output: <key', value'> pair, where the key' is the index of the cluster, value' is a string representing the new center.

1. Initialize each cluster (array) with the relevant sentences and score.

2. Initialize a counter NUM as 0 for each array to record the sum of sample number in the same cluster;
3. While (V.hasNext()){
4. Update the cluster centroid based on sentence scores obtained from each cluster.
5. For all the sentences in a cluster using all the features explained later, a new score is generated and aggregated with relevance score.
6. Using random sampling take a sample size *n* from NUM, from each cluster in the reducer and compute the mean.
7. Update the centroid with new mean.
8. Continue till convergence.
9. Best scored sentences can be selected from each cluster to generate summary for each aspect.
10. Output<key, value> pair;
11. End

The algorithm works as follows:

The input to the reduce function is the key which is the aspect word for a cluster and retrieved sentences for that aspect along with its relevant score. Here a counter is initialized to NUM to count the total number of samples for that cluster. Now the reducer computation is more focused on sentence score, which is computed based on the following important features to identify the important sentences used for generating a summary specific to an aspect. The features and their significance are detailed below:

- Word scoring

This feature comprises using word score based on each term in the sentence. Here each word receives a score, and the weight of each sentence is the sum of the scores from all the words. Here two components are used to calculate word score. One uses word frequency as a factor, and the other uses the familiar tf-idf as a score.

- Cue phrases

These cue phrases are considered to be good indicators of aspect-related opinions in the reviews. Some of the cue phrases are good, better, important, the best, significant, wonderful, awesome, and also some domain-specific terms like beautiful, large, clumsy, so on. The cue phrases are exclusively constructed, and the score is computed.

- Paraphrases

Paraphrases are important for a summary generation because they reveal whether two phrases convey the same meaning in different formats. This is more common in user reviews because many reviews convey the same information with different styles. We used WordNet, automatically constructed collections of near synonyms, which provide nominalizations of verbs and other derivationally related words across different POS categories. Also this can be used to match synonyms (e.g. 'to invent' and 'invention'), hypernyms–hyponyms, or, more generally, semantically related words across the two input expressions to identify paraphrases. We identified paraphrases within the cluster, and if it is found, the two sentences were fused, preserving its original meaning, and included in the summary. For the identification of paraphrases, we considered top-scored sentences with a threshold value of 10. So, redundancy is avoided, and also important meaningful sentences are included in the summary.

- Word co-occurrence

This feature is included to identify the important sequence of words which may contribute to summary. This is implemented using n-gram metric. Here 3 gram is used, assuming no user reviews can have more than three common terms co-occurring many times. High score is assigned to sentences with co-occurrence words appearing more in common. After summing up the score based on all the features, sentences are ordered based on decreasing order of scores.

4.3.4 Summary Generation

The top-most sentences from each cluster can be a summary for each aspect. Now even though the sentences with high score can be good representatives for a summary, some sentences which do not have a high score could also be a meaningful sentence to generate a summary. To take this into account, a random sampling technique to update the centroid of each cluster is included. Here random sampling technique is used to select the sentences within the cluster, and the mean for each cluster is computed. Also from all the reducers, the corresponding cluster mean values are merged, and the centroid for every iteration is computed using its average. Variation of the number of samples in the ensemble method makes the algorithm to perform best after minimum iterations. After updating the centroid for each cluster, the data is sent to mapper to verify the relevancy score. This step aids to retrieve and update the clusters and makes outliers not to be included in a cluster. The above method explained used reviews for generating summaries and used parallel k-means on its principle. This was found to be effective for feature-based text summary generation.

An Illustrative Example for MapReduce

Good Time: Movie Review

There is a saying that lightning never strikes the same place twice. Many of us, however, probably experienced a series of misfortunes at least once in our lives. When something bad happens, you think, 'Alright that should be it. Nothing worse can happen to me now' – and then it happens. But I guess few people have had a day (or night) as bad as Connie Nikas, the main character of an American criminal drama *Good Time*. Really, just how unfortunate can a man be?

Filmed in 2017 by Josh and Benny Safdie, the movie is their debut. Starring Robert Pattinson as Connie, the film not only reveals him as a brilliant actor but also redeems him of his role in *The Twilight Saga*. In other words, Connie looks totally different from what you would have expected of Pattinson, considering his previous characters. I will get to this aspect later; for now, let me briefly review the movie's plot. So, spoilers ahead – consider yourself warned.

Let's go. It starts with Connie breaking into a psychiatric clinic, where a psychiatrist interviews his brother Nick. Nick suffers from some kind of an autistic spectrum disorder, so he needs to go into therapy. Connie believes it only hurts his brother, so he interrupts one of such sessions and takes him away from the clinic.

Together, they rob a bank: as far as I understood, they need money to buy a farm where they can live all by themselves, of which they dream together. The brothers manage to escape with the money in a getaway car, but the bag with the money has an exploding dye pack in it as a safety measure to mark the robbers for the police. Nick and Connie have to make a stop at a restaurant to wash the dye away. Later, as they walk down the street, a police patrol officer stops them for an identification check; Nick panics and runs, causing the police to chase the two brothers. Connie flees, but Nick gets arrested.

Now, I have already mentioned Robert Pattinson's brilliant acting in this movie. From a glossy metrosexual vampire, he turned into a shabby, low-life 'white trash' guy. Yes, I do know he had other roles besides Edward Cullen. But just as there is always going to be a part of Harry Potter in Daniel Radcliffe, so will Edward be a part of Pattinson – at least for people of my generation. Anyways, in *Good Time*, there is nothing left from the 'old' Pattinson, so to say. A witty, risky, criminally inclined thug with a difficult life, he wants only to live a solitary life in peace with his brother – the only person he seems to truly care about. Bank robbery was perhaps the only way he could think of to get the money quickly.

The movie's pace (or rather, the suspense) quickly increases. It reminded me of one of my favorite movies, *The Talented Mr. Ripley*. Indeed, Connie and Ripley are somewhat alike. Both are nobodies; both virtuously manipulate other people in order to reach personal goals; both are skilled in improvisation and are unlikely to give up in difficult situations. And, just as in *The Talented Mr. Ripley*, Connie gradually pulls himself into bigger and bigger trouble.

Other characters are detailed and vivid as well. I felt sympathy for Nick. Unable to take full responsibility for his actions, he becomes a victim of his brother's ambition. I felt compassion towards Crystal and Corey – women Connie uses for his plans. I even felt pity for Ray, a slave to his addictions. These as well as other characters shown in the movie created an incredibly convincing picture of the bottom of American society and contributed to the atmosphere displayed by the film.

Acting and characters are not the only pillars the movie stands on. The visual images in *Good Time* are as important as the characters. A dull, gloomy picture filled with bluish and red neon lights perfectly conveys the mood of hopelessness, of endless days filled with poverty and the same routine. The greater part of the movie unfolds in the dark, on night streets, in urban slums. Connie's desire to get out of this dump becomes almost tangible, and his idea with the robbery starts looking like his only chance for a normal life. The camerawork always emphasizes the atmosphere in the film, and at some point, I found myself perceiving *Good Times* almost as a documentary.

And oh yes, the music. It is fantastic. The soundtrack was composed by Oneohtrix Point Never, a composer of experimental electronic music based in Massachusetts. After watching the film, I went directly to iTunes and downloaded his official *Good Time* soundtrack album – and I recommend you doing the same. In fact, it is so good, and it won the Soundtrack Award at the Cannes Festival 2017. You should rather listen to it yourself – or, which is better, watch the movie and hear the soundtrack alongside the story.

Overall, *Good Time* is a strong drama with complex characters, great visuals, and unforgettable music. It is a movie about criminals, but unlike most gangster movies, you will not find any romanticism or nostalgia here. It is a dark, hopeless underworld existing right next to our everyday lives – an underside of our society. Things depicted in *Good Time* happen every day – and this film is your peephole to the other side.

Consider the review: the stages in MapReduce are illustrated in the steps below:

Mapper stages:

Feature term as inputs to Mapper: actor, characters, visual images, music clusters formed in Mapper, and relevant sentences from Mapper:

i. Actor:
   a. Starring Robert Pattinson as Connie, the film not only reveals him as a **brilliant actor** but also **redeems him of his role** in *The Twilight Saga*. (20)
   b. Now, I have already mentioned Robert Pattinson's **brilliant acting** in this movie. (18)
   c. The acting and characters are not the only **pillars the movie** stands on. (15)

ii. Characters:
   a. Connie looks **totally different** from what you would have expected of Pattinson, considering his **previous characters**. (15)
   b. Other **characters are detailed and vivid** as well. (14)
   c. These as well as **other characters** shown in the movie created an **incredibly convincing** picture of the bottom of American society and contributed to the atmosphere displayed by the film. (20)
   d. The visual images in *Good Time* are as **important** as the **characters**. (12)

iii. Visual images
   a. The **visual images** in *Good Time* are as **important** as the characters. (20)
   b. Overall, *Good Time* is a strong drama with complex characters, **great visuals**, and unforgettable music. (18)
   c. A dull, **gloomy picture filled** with bluish and red neon lights perfectly conveys the mood of hopelessness, of endless days filled with poverty and the same routine. (18)
   d. The **camerawork always emphasizes** the atmosphere in the film, and at some point, I found myself perceiving *Good Times* almost as a documentary. (20)

iv. Music
   a. And oh yes, the **music**. It is **fantastic**. (20)
   b. The **soundtrack** was **composed** by Oneohtrix Point Never, a composer of **experimental electronic music** based in Massachusetts. (20)
   c. After watching the film, I went **directly to iTunes and downloaded** his official *Good Time* soundtrack album – and I **recommend you** doing the same. In fact, it is so **good**, and it won the Soundtrack Award at the Cannes Festival 2017. (20)
   d. You should rather **listen to it** yourself – or, which is **better, watch the movie** and **hear the soundtrack** alongside the story. (18)

These are the four clusters formed from mapper, and this forms the input to the reducer algorithm.

Based on the relevance score and centroid of the clusters, the sentences are selected for summary. In reducer, clusters are updated with new mean score of all sentences. The sentences which score nearest to centroid score are included in the cluster. Other sentences are rejected. Now the centroid for the four clusters is determined as:

i. Actor – 17.5
   a. Starring Robert Pattinson as Connie, the film not only reveals him not only as a **brilliant actor** but also **redeems him of his role** in *The Twilight Saga*. (20)
   b. Now, I have already mentioned Robert Pattinson's **brilliant acting** in this movie. (18)

ii. Characters – 15
   a. Connie looks **totally different** from what you would have expected of Pattinson, considering his **previous characters**. (15)
   b. These as well as **other characters** shown in the movie created an **incredibly convincing** picture of the bottom of the American society and contributed to the atmosphere displayed by the film. (20)

iii. Visual images – 19
   a. The **visual images** in *Good Time* are as **important** as the characters. (20)
   b. Overall, *Good Time* is a strong drama with complex characters, **great visuals**, and unforgettable music. (18)
   c. The **camerawork always emphasizes** the atmosphere in the film, and at some point, I found myself perceiving *Good Times* almost as a documentary. (20)

iv. Music – 19.5
   a. And oh yes, the **music**. It is **fantastic**. (20)
   b. The **soundtrack** was **composed** by Oneohtrix Point Never, a composer of **experimental electronic music** based in Massachusetts. (20)
   c. After watching the film, I went **directly to iTunes and downloaded** his official *Good Time* soundtrack album – and I **recommend you** doing the same. In fact, it is so **good**, and it won the Soundtrack Award at the Cannes Festival 2017. (20)

These are the sentences selected for summary for specific aspects. Again these are filtered based on feature scores as described above.

4.4 OTHER MR-BASED METHODS

Few other approaches that use MapReduce are briefed in this section. Nagwani (2015) [7] used a four-stage approach for document summarization. The corresponding framework is given in Figure 4.1.

The four phases involved are:

(i) Document clustering
(ii) Generation of cluster topics and terms using LDA
(iii) Generation of frequent terms
(iv) Sentence filtering and duplicate removal

All the stages consist of various MapReduce algorithms for performing the operations. These are explained in detail as follows:

i. Document clustering:

In order to perform clustering of the text documents, all the documents Di are brought together into one dataset, D. Then the k-means clustering algorithm is applied to perform the clustering of on the whole document set. K-clusters are generated. The set of clusters C = {C1, C2, …, CK}, where Ck(k = 1, 2, …, K) consists of group of similar documents belonging to a particular cluster Ci. Clustering ensures that similar set of text documents are grouped together and

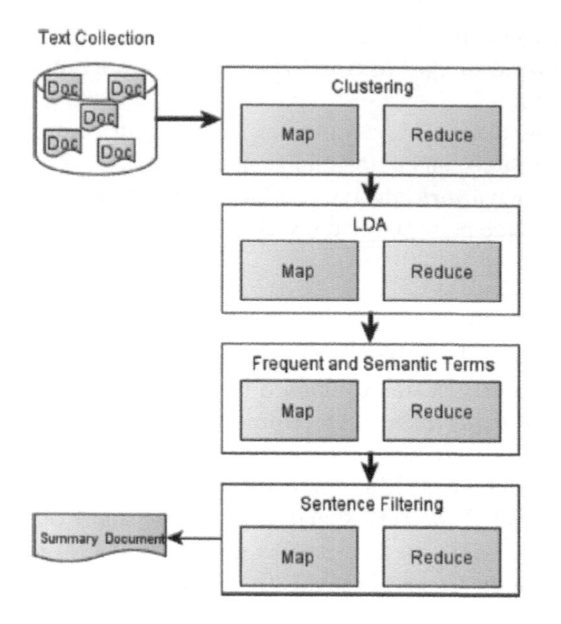

FIGURE 4.1 Multi-stage MapReduce framework for summarization

logically represents a theme (summarization unit) for effective summarization. The algorithm for document clustering in mapper and reducer phases is given in detail.

Mapper

- Initialize the cluster centers randomly and read them into memory from a sequence file.
- Iterate each cluster center for each input key–value pair. Compute Center (K,V).
- Calculate the distances and assign the nearest center with the lowest distance.
- Update the cluster center with its vector to the file system.

Reducer

- Iterate each value vector and calculate the mean.
- Update the new center from the calculated mean C.
- Check whether the cluster center is the same as the new center. If ( ), then stop, else go to the next step.
- If they are not equal, increment an update counter.

Clustering groups the documents based on similarity. The next step is to generate topics and terms for all the clusters using topic modeling technique.

ii. Generation of cluster topics and terms using LDA

The topic modeling technique is then applied on collective information to generate the topics from each text document clusters. Latent Dirichlet allocation (LDA) technique is used in this work for generating topics from each document cluster. The corresponding algorithms are given:

Mapper

```
For each cluster get the documents it contains and extract the
text collection from these documents.
       For each cluster Ci Ɛ { C1,C2,....,Cn}
              Extract the documents in Ci as { Di1,Di2,....,Dim}
              For each document extract and merge the text
                     from the text collection.
                     Text= Text U { Text Ɛ D }
                     Apply LDA topic modelling to these
                           collection and get the list of
                           topics for cluster
       Ci as Ti = {Ti1,Ti2......,Tik}.
              End for
       End for
End For
```

Reducer

```
Integrate the topics of all the clusters
For each cluster Ci Ɛ {C1,C2,.......CN}
       Extract the topics discovered by the LDA in the
               documents in Ci as
Ti = {Ti1,Ti2.......TiK}
       For each document extract the text and computer
               the text collection.
              Topics = Topics U {Tu Ɛ Ti}.
```

Topic and term identification has to be followed with semantic similar term identification. This is explained below:

iii. Generation of frequent terms

In the third stage, semantic similar terms are computed for each topic term generated in the previous stage. The semantic similar terms are generated over the MapReduce framework, and the generated semantic terms are added to the vector. Semantic similar term finding is an intensive computing operation. It requires going through the vocabulary and synonym data for the given term in the hierarchy of semantic relationship. MapReduce framework is utilized efficiently for handling this operation. The algorithm works as steps given below:

(i) The mapper computes the semantic similar terms for each topic term generated by the document cluster, and the reducer aggregates these terms and counts the frequencies of these terms (topic terms and semantic similar terms of topic terms) aggregately.
(ii) The mapper and reducer for semantic term generation from cluster topic terms are presented.
(iii) Then the terms are arranged in the descending order of frequency and top *N* topic terms (including the semantic similar terms) are selected.
(iv) These filtered terms are called as semantic similar frequent terms available in the document collection using the method ComputeSemanticSimilar(Ti) .
(v) The algorithm counts the number of occurrences of every word in a text collection.
(vi) Input key–values pairs take the form of (document id, doc) pairs stored on the distributed file system.
(vii) The key parameter is a unique identifier for the document, and the value parameter is the text of the document itself.
(viii) The mapper takes key–value pair as input, generates tokens from the document, and emits an intermediate key–value pair for every word.
(ix) The MapReduce execution makes sure that all values associated with the same key are brought together in the reducer.
(x) The final output of the algorithm is written to the distributed file system, one file per reducer.

Mapper

```
For each topic term in the topic list {T1,T2,.......TN}
      Get the semantic similar terms
      TSi = ComputeSemanticSimilar(Ti) // Pass the
            term Ti in WordNet API and extract the
            semantic similar terms in the sets TSi
            For all term t t Ɛ TSi present in the
                document D do
                      Emit (term t; count 1)
```

Reducer

```
For each term t, counts [c1,c2....]
Initialize the sum of term frequency as 0
```

```
For all count c Ɛ counts [c1,c2....]do
Update sum by adding count, i.e. sum+ = c
Emit (term t, count sum)
```

Now the identified similar terms and topics are used for sentence filtering in the next stage, to generate summary sentences.

iv. Sentence filtering and duplicate removal

In the last stage, the original text document collection is distributed over the mappers, and using parsing techniques, sentences are extracted from individual document by the mappers. The sentences which consist of the frequent terms and their semantic similar terms are filtered from the original text collection and added to the summary document (in other words the filtered terms participate in the summary document). The final summary is generated after traversing all the documents in the document collections. The corresponding pseudo codes for the algorithms are presented.

Mapper

```
Select one document at a time from the document collection
For each Document D Ɛ {D1,D2,......}
       Extract the sentences from Document D as { Si1,Si2,.....}
             If Sik contains the terms present in TSi
                       Filter the sentences Sik containing the
                              terms and add it to Vector .
                       Vector = Vector U {SK}
```

Reducer

a. Integrate all the filtered sentences and produce a single document presenting the summary.
b. Summary = Summary U {Sik}

The dataset is considered as legal cases from UC Irvine (UCI) machine learning repository. Since four stages are used with MapReduce framework, the system provides an efficient summary of documents. The subsequent work using MapReduce was published by Kitsos et al. (2014) for entity-based summarization. This work consists of two phases: First, the output of the analysis over the full search results can be considered a summarization task over text data appropriate for exploration by human users. Second, the implementations perform a first summarization pass over the full search results to (i) analyze a small sample of the documents and provide the end-user a quick preview of the complete analysis;

and (ii) collect the sizes of all files and use them in the second (full) pass to better partition that dataset achieving better load balancing. There are many works proposed by Mackey and Israel 2018; Jimmy and Dyer 2010; Braka and Sulaiman 2017.

This provides a scalable method for entity-based summarization of Web search results at query time using the MapReduce programing framework. Performing entity mining over the full contents of results provided by keyword search systems require downloading and analyzing those contents at query time. Using a single machine to perform such a task for queries returning several thousands of hits becomes infeasible due to its high computational and memory cost. So, a sequential Named Entity Mining algorithm is decomposed into an equivalent distributed MapReduce algorithm and deployed on the Amazon EC2 Cloud. Best load balancing was obtained using (maximizing utilization of resources) two MapReduce procedures. Their scalability and the overall performance under different configuration/tuning parameters in the underlying platform (Apache Hadoop) were analyzed. Utilizing Amazon EC2 VMs for analyzing 300 MB datasets, for a total runtime of less than 7 ms, achieved better performance.

The next work by Belereo and Chaudhari (2017) was designed using DBSCAN algorithm which works with MapReduce framework for clustering and Hidden Markov Model for summarization. The summarization process is performed in three main stages and provides a modular implementation of large number of documents for summarization.

To generate the clusters of sentences from the document set, DBSCAN algorithm is used. To group data points together, DBSCAN algorithm uses similarity metric, usually in the form of a distance. Two inputs are required for DBSCAN, that is, *c* which is the minimum number of points in a dense cluster and distance. Every data point is visited by DBSCAN in the dataset, and it draws a radius around the point. If there is at least *c* number in the radius, it is called as dense point. If there is no minimum point in the radius, but there is dense point, it is called as border point. If no points are found, then it is called noise point. MapReduce framework is used for preprocessing of data and clustering. Mapper is responsible for computation of the semantic similar words, and reducer aggregates words and counts the number of occurrences of these words. Clusters of the similar words are generated from preprocessing phase. Filtered words are distributed in various clusters using mappers. The sentences which are similar and used frequently in document are filtered from the document set and added to the summary document. The final summary is generated after filtering all the documents from a collection of various documents.

Cognitive Models for Parallel Tasks:

The paper proposed by Zamuda and Lloret (2020) presents the tackling of a hard optimization problem of computational linguistics, specifically automatic multi-document text summarization, using grid computing. The main challenge of multi-document summarization is to extract the most relevant and unique information effectively and efficiently from a set of topic-related documents, constrained to a specified length. In the Big Data/Text era, where the information increases exponentially, optimization becomes essential in the selection of the most representative sentences for generating the best summaries. Therefore, a data-driven summarization model is proposed and optimized during a run of differential evolution (DE).

Different DE runs are distributed to a grid in parallel as optimization tasks, seeking high processing throughput despite the demanding complexity of the linguistic model, especially on longer multi-documents where DE improves results given more iterations. This model can be optimized using cognitive features or models in all stages to improve the summarization process.

4.5 SUMMARY

Parallel processing methods based on both computation and data had been discussed in this chapter. Since algorithmic techniques for text document summarization have been found minimum, distributed frameworks such as MapReduce have been widely used. Further the summarization systems processed using MapReduce algorithms have been found to produce more efficient results than traditional algorithms.

4.6 EXAMPLES

K-Means Clustering Using MapReduce

```
Step 1: Input: P1-P16
4 Input splits: Split 0 to Split 3
```

This is represented in Figure 4.2.

Now assign centroids for the clusters C0–C4

Now in the reduce phase, and based on the output of the combiner, you need to recalculate the centroids by iterating over the values and output the intermediate centroids.

Parallel LDA Example (Using Gensim Package)

Represented in Figure 4.3.

Only 100 papers are considered: Screenshot in Figure 4.4 after retaining only 100 papers.

Word cloud for most common words: In Figure 4.5.

Building LDA model:

Build a model with ten topics where each topic is a combination of keywords, and each keyword contributes a certain weightage to the topic.

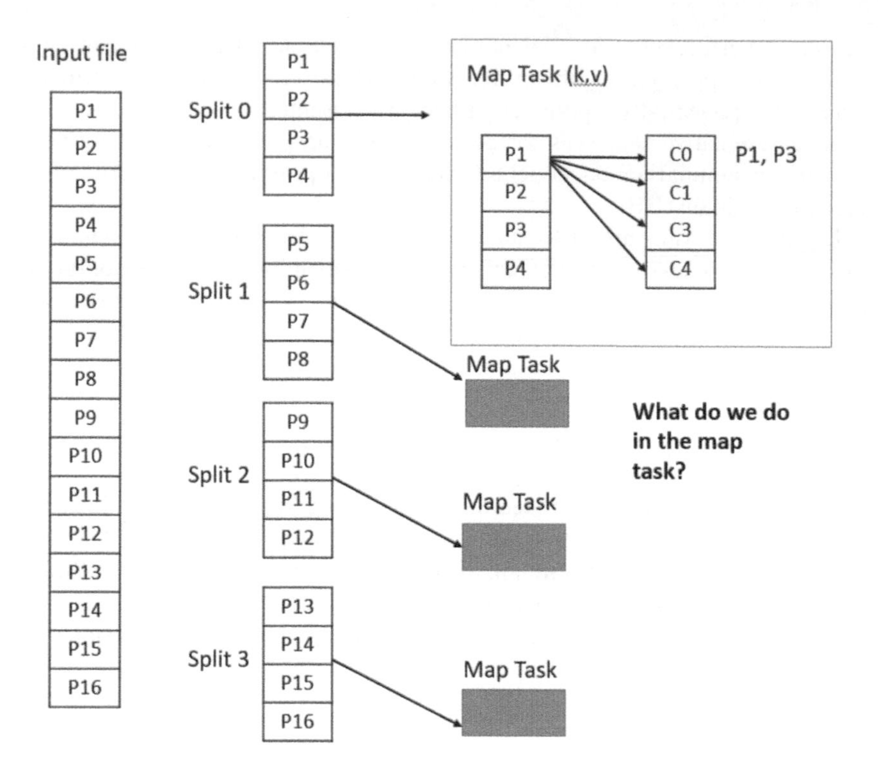

FIGURE 4.2 K-means clustering using MapReduce

Out[1]:

	id	year	title	event_type	pdf_name	abstract	paper_text
0	1	1987	Self-Organization of Associative Database and ...	NaN	1-self-organization-of-associative-database-an...	Abstract Missing	767\n\nSELF-ORGANIZATION OF ASSOCIATIVE DATABA...
1	10	1987	A Mean Field Theory of Layer IV of Visual Cort...	NaN	10-a-mean-field-theory-of-layer-iv-of-visual-c...	Abstract Missing	683\n\nA MEAN FIELD THEORY OF LAYER IV OF VISU...
2	100	1988	Storing Covariance by the Associative Long-Ter...	NaN	100-storing-covariance-by-the-associative-long...	Abstract Missing	394\n\nSTORING COVARIANCE BY THE ASSOCIATIVE\n...
3	1000	1994	Bayesian Query Construction for Neural Network...	NaN	1000-bayesian-query-construction-for-neural-ne...	Abstract Missing	Bayesian Query Construction for Neural\nNetwor...
4	1001	1994	Neural Network Ensembles, Cross Validation, an...	NaN	1001-neural-network-ensembles-cross-validation...	Abstract Missing	Neural Network Ensembles, Cross\nValidation, a...

FIGURE 4.3 Screenshot for parallel LDA example

Out[2]:

	year	title	abstract	paper_text
0	1987	Self-Organization of Associative Database and ...	Abstract Missing	767\n\nSELF-ORGANIZATION OF ASSOCIATIVE DATABA...
1	1987	A Mean Field Theory of Layer IV of Visual Cort...	Abstract Missing	683\n\nA MEAN FIELD THEORY OF LAYER IV OF VISU...
2	1988	Storing Covariance by the Associative Long-Ter...	Abstract Missing	394\n\nSTORING COVARIANCE BY THE ASSOCIATIVE\n...
3	1994	Bayesian Query Construction for Neural Network...	Abstract Missing	Bayesian Query Construction for Neural\nNetwor...
4	1994	Neural Network Ensembles, Cross Validation, an...	Abstract Missing	Neural Network Ensembles, Cross\nValidation, a...

FIGURE 4.4 Screenshot after considering only 100 papers

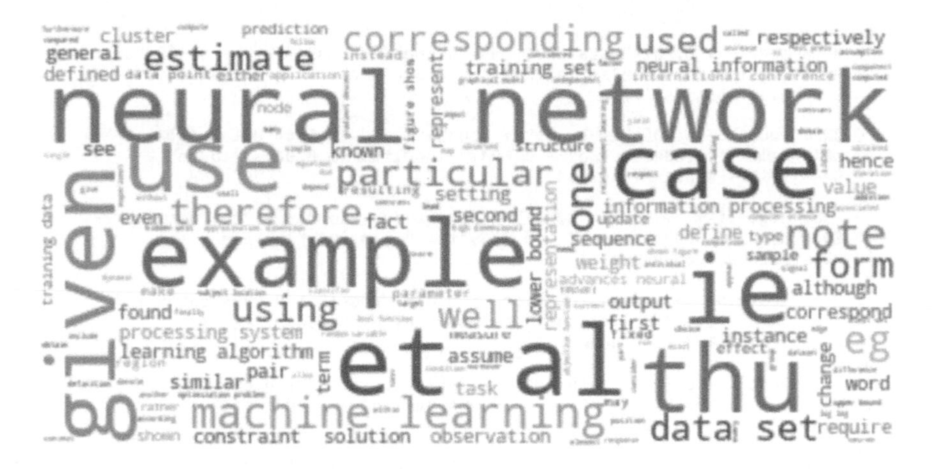

FIGURE 4.5 Word cloud for the most common words

Sample Code: (Using Gensim Package)

```
from pprint import pprint
# number of topics
num_topics = 10
# Build LDA model
lda_model = gensim.models.LdaMulticore(corpus=corpus,
id2word=id2word,num_topics=num_topics)
# Print the Keyword in the 10 topics
pprint(lda_model.print_topics())
doc_lda = lda_model[corpus]
```

After running this output as in Figure 4.6 we can visualize the result using any visualization package.

Example: Creating an Inverted Index

An inverted index consists of a list of all the unique words that appear in any document, and for each word, a list of the documents in which it appears. The inverted index is useful for the fast retrieval of relevant information.

Let's look at building an inverted index for a set of tweets based on their hashtags and how we can map the solution as a MapReduce.

Input Data:

'It's not too late to vote. #ElectionDay'
'Midtown polling office seeing a steady flow of voters! #PrimaryDay'
'Today's the day. Be a voter! #ElectionDay'
'Happy #PrimaryDay'
'Say NO to corruption & vote! #ElectionDay'
'About to go cast my vote…first time #ElectionDay'

```
[(0,
  '0.008*"model" + 0.007*"data" + 0.005*"using" + 0.005*"set" + 0.005*"one" + '
  '0.004*"algorithm" + 0.004*"learning" + 0.004*"models" + 0.004*"time" + '
  '0.003*"distribution"'),
 (1,
  '0.006*"model" + 0.005*"learning" + 0.005*"two" + 0.004*"data" + '
  '0.004*"time" + 0.004*"using" + 0.004*"algorithm" + 0.004*"figure" + '
  '0.004*"set" + 0.004*"function"'),
 (2,
  '0.006*"algorithm" + 0.006*"model" + 0.005*"function" + 0.005*"network" + '
  '0.005*"data" + 0.004*"using" + 0.004*"time" + 0.004*"learning" + '
  '0.004*"one" + 0.003*"set"'),
 (3,
  '0.006*"learning" + 0.006*"function" + 0.005*"model" + 0.005*"data" + '
  '0.005*"one" + 0.004*"algorithm" + 0.004*"set" + 0.004*"time" + '
  '0.004*"using" + 0.003*"number"'),
 (4,
  '0.007*"model" + 0.007*"learning" + 0.006*"algorithm" + 0.005*"set" + '
  '0.005*"data" + 0.004*"function" + 0.004*"using" + 0.004*"one" + '
  '0.004*"figure" + 0.003*"time"'),
 (5,
  '0.008*"model" + 0.006*"algorithm" + 0.005*"data" + 0.005*"set" + '
  '0.004*"function" + 0.004*"learning" + 0.004*"one" + 0.004*"used" + '
  '0.003*"time" + 0.003*"also"'),
 (6,
  '0.005*"data" + 0.005*"function" + 0.004*"model" + 0.004*"algorithm" + '
  '0.004*"learning" + 0.004*"using" + 0.004*"figure" + 0.004*"problem" + '
  '0.003*"training" + 0.003*"two"'),
 (7,
  '0.009*"learning" + 0.007*"model" + 0.007*"data" + 0.005*"set" + '
  '0.005*"network" + 0.004*"one" + 0.004*"algorithm" + 0.004*"number" + '
  '0.004*"using" + 0.003*"log"'),
 (8,
  '0.009*"learning" + 0.006*"data" + 0.005*"algorithm" + 0.004*"function" + '
  '0.004*"problem" + 0.004*"set" + 0.004*"using" + 0.004*"time" + 0.004*"two" '
  '+ 0.003*"model"'),
 (9,
  '0.006*"model" + 0.005*"data" + 0.004*"learning" + 0.004*"one" + 0.004*"set" '
  '+ 0.004*"two" + 0.004*"algorithm" + 0.004*"number" + 0.003*"problem" + '
  '0.003*"function"')]
```

FIGURE 4.6 Execution of code

MapReduce Mapping:

map : *tweet*→ (*hastag, tweet*)
reduce:(*hastag* ,{*list*(*tweet*)})→ *hastag*,{*list*(*tweet*)}

Map Output:

("ElectionDay", "It's not too late to vote. #ElectionDay")
("PrimaryDay", "Midtown polling office seeing a steady flow of voters! #PrimaryDay")
("ElectionDay", "Today's the day. Be a voter! #ElectionDay ")
("PrimaryDay", "Happy #PrimaryDay")
("ElectionDay", "Say NO to corruption & vote! #ElectionDay")
("ElectionDay", "About to go cast my vote...first time #ElectionDay")

Reduce Input:

Reducer 1:
("ElectionDay", "It's not too late to vote. #ElectionDay")
("ElectionDay", "Today's the day. Be a voter! #ElectionDay ")
("ElectionDay", "Say NO to corruption & vote! #ElectionDay")
("ElectionDay", "About to go cast my vote...first time #ElectionDay")
Reducer 2:
("PrimaryDay", "Midtown polling office seeing a steady flow of voters! #PrimaryDay")
("PrimaryDay", "Happy #PrimaryDay")
Reduce Output:
("ElectionDay", ["It's not too late to vote. #ElectionDay",
"Today's the day. Be a voter! #ElectionDay ",
"Say NO to corruption & vote! #ElectionDay",
"About to go cast my vote...first time #ElectionDay"])
("PrimaryDay", ["Midtown polling office seeing a steady flow of voters! #PrimaryDay",
"Happy #PrimaryDay"])

Example: Relational Algebra (Table JOIN)

MapReduce can be used to join two database tables based on common criteria. Let's take an example. We have two tables, where the first contains an employee's personal information primarily keyed on SSN and the second table includes the employee's income again keyed on SSN. We would like to compute the average income in each city in 2016.

This computation requires a JOIN operation on these two tables. We will map the problem to a two-phase MapReduce solution. The first phase effectively creates a JOIN on the two tables using two map functions (one for each table), and the second phase gathers the relevant data for calculating the desired statistics.

Input Data:
Table 4.1 (SSN, {Personal Information})

111222:(Stephen King; Sacramento, CA)
333444:(Edward Lee; San Diego, CA)
555666:(Karen Taylor; San Diego, CA)

Table 4.2 (SSN, {year, income})

111222:(2016,$70000),(2015,$65000),(2014,$6000),...
333444:(2016,$72000),(2015,$70000),(2014,$6000),...
555666:(2016,$80000),(2015,$85000),(2014,$7500),...

MapReduce Mapping:

Stage 1 (Table JOIN)
*maptable*1 : *recordtable*1 → (*SSN*, *city*)
*maptable*2 : *recordtable*2 → *SSN*, 2016)
reduce: (*SSN*, {*City*,*Income*2016}) → (*SSN*,(*City*,*Income*2016)
Stage 2
map: *SSN*, (*City*,2016) → (*City*,*Income*2016)
reduce: *City*, {(*list*(*Income*2016)} → *City*,*avg*(*Income*2016)

Stage 1
Map Output:

Mapper 1a: (SSN, city)
(111222, "Sacramento, CA")
(333444, "San Diego, CA)
(555666, "San Diego, CA)
Mapper 1b: (SSN, income 2016)
(111222, $70000)
(333444, $72000)
(555666, $80000)
Reduce Input: (SSN, city), (SSN, income)
(111222, "Sacramento, CA")
(111222, $70000)
(333444, "San Diego, CA")
(333444, $72000)
(555666, "San Diego, CA")
(555666, $80000)
Reduce Output: (SSN, [city, income])
(111222, ["Sacramento, CA", 70000])
(333444, ["San Diego, CA", 72000])
(555666, ["San Diego, CA", 80000])

Stage 2:

Map Input: (SSN, [city, income])
(111222, ["Sacramento, CA", 70000])
(333444, ["San Diego, CA", 72000])
(555666, ["San Diego, CA", 80000])
Map Output: (city, income)
("Sacramento, CA", 70000)
("San Diego, CA", 72000)
("San Diego, CA", 80000)
Reduce Input: (city, income)
Reducer 2a:

("Sacramento, CA", 70000)

Reducer 2b:
("San Diego, CA", 72000)
("San Diego, CA", 80000)
Reduce Output: (city, average [income])
Reducer 2a:

("Sacramento, CA", 70000)

Reducer 2b:
("San Diego, CA", 76000)

2. How the solution will differ if the employee is allowed to have multiple addresses, i.e. there can be multiple addresses per SSN.
3. Give the output for the following input text using MapReduce at different stages:

Input Text:

This is a cat
Cat sits on a roof
The roof is a tin roof
There is a tin can on the roof
Cat kicks the can
It rolls on the roof and falls on the next roof
The cat rolls too
It sits on the can

4. Suppose the Indian government has assigned you the task to count the population of India. You can demand all the resources you want, but you have to do this task in 4 months. Calculating the population of such a large country is not an easy task for a single person (you). So MapReduce approach to solve the problem?

REFERENCES

Baraka, R. S., and Al Breem, S. N. 2017. "Automatic arabic text summarization for large scale multiple documents using genetic algorithm and mapreduce." In *2017 Palestinian International Conference on Information and Communication Technology (PICICT)* (pp. 40–45). Gaza, Palestine: IEEE.

Belerao, K. T., and Chaudhari, S. B. 2017. "Summarization using mapreduce framework based big data and hybrid algorithm (hmm and dbscan)." In *2017 IEEE International Conference on Power, Control, Signals and Instrumentation Engineering (ICPCSI)* (pp. 377–380). Chennai, India: IEEE.

Kitsos, I., Magoutis, K., and Tzitzikas, Y. 2014. "Scalable entity-based summarization of web search results using MapReduce." *Distributed and Parallel Databases* 32(3): 405–446.

Lin, J., and Dyer, C. 2010. Data-intensive text processing with MapReduce. *Synthesis Lectures on Human Language Technologies* 3(1): 1–177. Springer Nature Switzerland.
Mackey, A., and Cuevas, I. 2018. "Automatic text summarization within big data frameworks." *Journal of Computing Sciences in Colleges* 33(5): 26–32.
Maria, K. A., Jaber, K. M., and Ibrahim, M. N. "A new model for arabic multi-document text summarisation." *International Journal of Innovative Computing, Information and Control* 14(4): 1443–1452.
Meena, Y. K., and Gopalani, D. 2015. "Evolutionary algorithms for extractive automatic text summarization." *Procedia Computer Science* 48(Supplement C): 244–249.
Nagwani, N. K. 2015. "Summarizing large text collection using topic modeling and clustering based on MapReduce framework." *Journal of Big Data* 2(1): 6.
Priya, V., and Umamaheswari, K. 2016. "Ensemble based parallel k means using map reduce for aspect based summarization." In *Proceedings of the International Conference on Informatics and Analytics*, Pondhicherry,India: ACM (pp. 1–9).
Wang, M., and Tang, X. 2012, May. "Extract summarization using concept-obtained and hybrid parallel genetic algorithm." In *2012 8th International Conference on Natural Computation*, Chongqing, China: IEEE (pp. 662–664).
Zamuda, A., and Lloret, E. 2020. "Optimizing data-driven models for summarization as parallel tasks." *Journal of Computational Science* 42: 101101.

SAMPLE CODE

```
import org.apache.commons.io.FileUtils;
import org.apache.hadoop.conf.Configuration;
import org.apache.hadoop.conf.Configured;
import org.apache.hadoop.fs.Path;
import org.apache.hadoop.fs.RemoteIterator;
import org.apache.hadoop.io.DoubleWritable;
import org.apache.hadoop.io.IntWritable;
import org.apache.hadoop.io.SequenceFile;
import org.apache.hadoop.io.Text;
import org.apache.hadoop.mapreduce.Job;
import org.apache.hadoop.mapreduce.lib.input.FileInputFormat;
import org.apache.hadoop.mapreduce.lib.output.FileOutputFormat;
import org.apache.hadoop.mapreduce.lib.output.TextOutputFormat;
import org.apache.hadoop.mapreduce.lib.input.SequenceFileInputFormat;
import org.apache.hadoop.mapreduce.lib.output.SequenceFileOutputFormat;
import org.apache.hadoop.fs.FileStatus;
import org.apache.hadoop.fs.FileSystem;
import org.apache.hadoop.fs.FileUtil;
import org.apache.hadoop.fs.LocatedFileStatus;
import org.apache.hadoop.util.Tool;
import org.apache.hadoop.util.ToolRunner;

import java.io.BufferedReader;
import java.io.BufferedWriter;
import java.io.File;
import java.io.FileInputStream;
import java.io.FileWriter;
import java.io.IOException;
import java.io.InputStreamReader;
import java.io.OutputStreamWriter;
import java.io.PrintWriter;
import java.io.Reader;
import java.io.StringReader;
import java.util.ArrayList;
import java.util.Formatter;
import java.util.Iterator;
import java.util.List;
import java.util.Locale;
import java.util.Random;
import java.util.Scanner;
```

```
import java.util.TreeSet;
import java.util.regex.Matcher;
import java.util.regex.Pattern;
import org.apache.log4j.Logger;
import cc.mallet.pipe.CharSequence2TokenSequence;
import cc.mallet.pipe.CharSequenceLowercase;
import cc.mallet.pipe.Pipe;
import cc.mallet.pipe.SerialPipes;
import cc.mallet.pipe.TokenSequence2FeatureSequence;
import cc.mallet.pipe.TokenSequenceRemoveStopwords;
import cc.mallet.pipe.iterator.CsvIterator;
import cc.mallet.topics.ParallelTopicModel;
import cc.mallet.topics.TopicInferencer;
import cc.mallet.types.Alphabet;
import cc.mallet.types.FeatureSequence;
import cc.mallet.types.IDSorter;
import cc.mallet.types.Instance;
import cc.mallet.types.InstanceList;
import cc.mallet.types.LabelSequence;
/*
 * Text Summarization Driver to initiate summarization of multiple documents
 * in the collection by the following phases:
 * Phase1: Preprocessing and vector representation of documents
 * Phase2: KMeans clustering.
 * Phase3: LDA Topic Modelling.
 * Phase4: Semantic term generation.
 * Phase5: Summarization.
 * */
public class TextSummarizationDriver extends Configured implements Tool {
        private static final Logger LOG = Logger.getLogger(TextSummarizationDriver.class);
        /*
         * The main method invokes the Text Summarization Driver ToolRunner, which
         * creates and runs a new instance of Text Summarization job.
         */
        public static void main(String[] args) throws Exception {
                Configuration configuration = new Configuration();
                System.exit(ToolRunner.run(configuration, new TextSummarizationDriver(),
                        args));
        }
        @Override
        public int run(String[] args) throws Exception {
                FileSystem fs = null;
                Configuration config = new Configuration();
                try {
                      System.out.println("*************PREPROCESS
                                      START****************");
                      fs = FileSystem.get(config);
                      FileStatus[] status = fs.listStatus(new Path(args[0])); // input
                                      // path
                                      // containing
                                      // file
                                      // collection
                      for (int i = 0; i < status.length; i++) {
                              BufferedReader br = new BufferedReader(new InputStrea
                                         mReader(fs.open(status[i].getPath())));
                              String lineRead;
                              lineRead = br.readLine();
                              String doc = "";
                              boolean value = false;
                              while (lineRead != null) {
                                      if (value) {
                                              doc += lineRead;
                                      }
                                      value = true;
                                      lineRead = br.readLine();
                              }
                              String[] linesArray = doc.split("\\.");
                              Path pt2 = new Path(args[1] + "/file" + i + ".txt"); //
                                  output
                                      // file
                                      // path
                      BufferedWriter bw = new BufferedWriter(new OutputStreamWriter(fs
                                  .create(pt2, true)));
                      for (String line : linesArray) {
```

```
                bw.write(line + ".");
                bw.newLine();
        }
        bw.close();
        }
} catch (Exception e) {
        e.printStackTrace();
}
System.out.println("*************PREPROCESS ENDS****************");
System.out.println("*************VECTOR CREATION STARTS****************");
boolean isClustered = false;
int res_termfrequency = ToolRunner.run(new TermFrequency(), args);
if (res_termfrequency != 0) {
        return 1;
}
int res_tfidf = ToolRunner.run(new TFIDF(), args);
if (res_tfidf != 0)
        return 1;
System.out.println("*************VECTOR CREATION ENDS****************");
System.out.println("************* K-MEANS STARTS****************");
isClustered = runKMeans(args, fs, config);
if (!isClustered)
        return 1;
   System.out.println("************* K-MEANS ENDS****************");
   System.out.println("************* LDA STARTS****************");
// call LDA
boolean performLDA = performLDA(args);
if (!performLDA) {
        return 1;
}
// Combine all topics from LDA Phase
boolean combineTopics = combineTopics(args[11], args[6]);
if (!combineTopics) {
        return 1;
}
System.out.println("************* LDA ENDS ****************");
System.out.println("************* SEMANTIC TERM STARTS****************");
System.setProperty("wordnet.database.dir", args[9]);
boolean createdSemanticTerms = createSemanticTerms(args[6], args[12]);
if (!createdSemanticTerms) {
        return 1;
}
boolean isSorted = cleanAndSorting(args[12], args[7]);
if (!isSorted) {
        return 1;
}
System.out.println("************* SEMANTIC TERM ENDS****************");
System.out.println("************* SUMMARIZATION STARTS****************");
int res_summarizer = ToolRunner.run(new Summarizer(), args);
if (res_summarizer != 0) {
        return 1;
}
System.out.println("************* SUMMARIZATION ENDS****************");
return 0;
}
/*
 * Triggers KMeans algorithm
 */
private boolean runKMeans(String[] args, FileSystem fs, Configuration conf)
                throws IOException, ClassNotFoundException,
                        InterruptedException {
        int iteration = 1;
        conf.set("num.iteration", iteration + "");
        Path in = new Path(args[4]);
        Path center = new Path(args[13]);
        conf.set("centroid.path", center.toString());
        Path out = new Path(args[14]);
        Job job = Job.getInstance(conf);
        job.setJobName("KMeans Clustering");
        job.setMapperClass(KMeansMapper.class);
job.setReducerClass(KMeansReducer.class);
job.setJarByClass(KMeansMapper.class);
FileInputFormat.addInputPath(job, in);
System.out.println("****** WRITING INITIAL CENTROIDS*********");
```

```
                writeCenters(conf, center, fs, args);
                System.out.println("****** FINISHED WRITING CENTROIDS*********");
                FileOutputFormat.setOutputPath(job, out);
                job.setOutputKeyClass(ClusterCenter.class);
                job.setOutputValueClass(Text.class);
                job.waitForCompletion(true);
                long counter = job.getCounters().findCounter(KMeansReducer.Counter.CON
                               VERGED).getValue();
                iteration++;
                int iter = 0;
                while (iter < 100) {
                       conf = new Configuration();
                       conf.set("centroid.path", center.toString());
                       conf.set("num.iteration", iteration + "");
                       job = Job.getInstance(conf);
                       job.setJobName("KMeans Clustering " + iteration);
                       job.setMapperClass(KMeansMapper.class);
                       job.setReducerClass(KMeansReducer.class);
                       job.setJarByClass(KMeansMapper.class);
                       in = new Path(args[15] + (iteration - 1) + "/");
                       out = new Path(args[15] + iteration);
                       FileInputFormat.addInputPath(job, in);
                       FileOutputFormat.setOutputPath(job, out);
                       job.setOutputKeyClass(ClusterCenter.class);
                       job.setOutputValueClass(Text.class);
                       job.waitForCompletion(true);
                       iteration++;
                       counter = job.getCounters().findCounter(KMeansReducer.Counter.CON
                                     VERGED).getValue();
                       iter++;
                }
                Path result = new Path(args[15] + (iteration - 1) + "/");
                FileStatus[] stati = fs.listStatus(result);
                for (FileStatus status : stati) {
                       if (!status.isDirectory()) {
                              Path path = status.getPath();
                              if (!path.getName().equals("_SUCCESS")) {
                                     conf = new Configuration();
                                     job = Job.getInstance(conf);
                                     job.setJobName("KMeans Clustering Final");
                                     job.setMapperClass(KMeansMapperFinal.class);
                                     job.setReducerClass(KMeansReducerFinal.class);
                                     job.setJarByClass(KMeansMapper.class);
                                     out = new Path(args[16]); // intermediate
                                     FileInputFormat.addInputPath(job, result);
                                     FileOutputFormat.setOutputPath(job, out);
                                     job.setOutputKeyClass(Text.class);
                                     job.setOutputValueClass(Text.class);
                                     if (job.waitForCompletion(true)) {
                                            try {
                                                   Path kMeanFiles = new
                                                        Path(args[17]);
                                                   BufferedReader br = new
BufferedReader(new InputStreamReader(fs.open(kMeanFiles)));
                                                   String line;
                                                   line = br.readLine();
                                                   int i = 0;
                                                   while (line != null) {
                                                          String[] files = line.spl
                                                                 it("\t");
                                                   String[] fileNames = files[1].
                                                                    split(",");
                                                   String clus = "Cluster" + i;
                                                   Path newFolderPath = new
                                                        Path(args[18] + clus);
                                                   fs.mkdirs(newFolderPath); //
                                                      Create new
                              // Directory
                                                   for (String fileName : fileNames) {
                                                          File localFilePath = new
                                                          File(args[1] + "/" +
                                                          fileName);
                                                          File hdfsFilePath = new
                                                          File(newFolderPath + "/"
                                                          + fileName);
```

```
                                                FileUtil.copy(fs, new
                                                    Path(args[1] + "/" +
                                                    fileName), fs,
                                                new Path(newFolderPath +
                                                    "/"), false, conf);
                                            }
                                            line = br.readLine();
                                            i++;
                                        }
                            } catch (Exception e) {
                                        System.out.println(e);
                            }
                        }
                    }
                }
        }
        return true;
}
/*
 * Writing cluster centers to the file system
 */
@SuppressWarnings("deprecation")
public static void writeCenters(Configuration conf, Path center, FileSystem fs, String[]
        args) throws IOException {
        try (SequenceFile.Writer centerWriter = SequenceFile.createWriter(fs, conf,
                center, ClusterCenter.class,
                                IntWritable.class)) {
                        final IntWritable value = new IntWritable(0);
                        System.out.println("Enter the number of clusters");
                        Scanner sc = new Scanner(System.in);
                        int noOfClusters = sc.nextInt();
                        String dataFilePath = args[4];
                        Path dataPath = new Path(dataFilePath + "/part-r-00000");
                        BufferedReader br = new BufferedReader(new InputStreamReader(fs.op
                            en(dataPath)));
                        String line;
                        line = br.readLine();
                        Pattern p = Pattern.compile("\\[(.*?)\\]");
                        List<String> lines = new ArrayList<String>();
                        while (line != null) {
                        lines.add(line);
                        line = br.readLine();
                        }
                        for (int i = 0; i < noOfClusters; i++) {
                                Random r = new Random();
                                String randomLine = lines.get(r.nextInt(lines.size()));
                                String[] fileNameVectors = randomLine.split("=");
                                Matcher m = p.matcher(fileNameVectors[1]);
                                String v = null;
                                while (m.find()) {
                                        v = m.group(1);
                                }
                                String[] vec = v.split(",");
                                double[] vecArray = new double[vec.length];
                                for (int j = 0; j < vec.length; j++) {
                                        String trim = vec[j].replaceAll("\\s+", "");
                                        vecArray[j] = Double.parseDouble(trim);
                                }
                                VectorWritable vw = new VectorWritable(vecArray);
                                centerWriter.append(new ClusterCenter(vw), value);
                        }
                }
        }

        // Performing LDA
        private boolean performLDA(String[] args) throws IOException,
ClassNotFoundException, InterruptedException {
                // TODO Auto-generated method stub

                Configuration conf = new Configuration();
                FileSystem fs = FileSystem.get(conf);
                Boolean jobStatus = true;
                Path result = new Path(args[5]);
                int i = 0;
```

```
        FileStatus[] stati = fs.listStatus(result);
        for (FileStatus status : stati) {
              if (status.isDirectory() && jobStatus) {
              Job job = Job.getInstance(getConf(), "performLDA");
                      job.setJarByClass(this.getClass());
                      Path path = status.getPath();
                      FileInputFormat.addInputPath(job, path);
                      FileOutputFormat.setOutputPath(job, new Path(args[19] + i));
                      job.setMapOutputKeyClass(Text.class);
                      job.setMapOutputValueClass(Text.class);
                      job.setMapperClass(LDAMapper.class);
                      job.setOutputKeyClass(Text.class);
                      job.setOutputValueClass(IntWritable.class);
                      job.setReducerClass(LDAReducer.class);
                      jobStatus = job.waitForCompletion(true);
                      i = i + 1;
              }
        }
        return jobStatus;
}

// combining similar topics
public boolean combineTopics(String input, String output)
              throws IOException, ClassNotFoundException, InterruptedException {
Configuration conf = new Configuration();
FileSystem fs = FileSystem.get(conf);
Path result = new Path(input);
BufferedWriter writer = new BufferedWriter(new OutputStreamWriter(fs.create(new
                    Path(output), true)));
FileStatus[] stati = fs.listStatus(result);
for (FileStatus status : stati) {
        if (status.isDirectory()) {
              FileStatus[] stati1 = fs.listStatus(status.getPath());
              for (FileStatus status1 : stati1) {
                      if (!status1.getPath().getName().equals("_SUCCESS")) {
                      BufferedReader br = new BufferedReader(new InputStrea
                                     mReader(fs.open(status1.getPath())));
                      String line;
                      line = br.readLine();
                      while (line != null) {
                              writer.write(line.trim() + "\n");
                              line = br.readLine();
                      }
              }
        }
        }
}
writer.close();
return true;
}

// Creating semantic terms
private boolean createSemanticTerms(String input, String output)
        throws IOException, ClassNotFoundException, InterruptedException {
Job job = Job.getInstance(getConf(), "createSemanticTerms");
job.setJarByClass(this.getClass());
        FileInputFormat.addInputPaths(job, input);
        FileOutputFormat.setOutputPath(job, new Path(output));
        job.setMapperClass(SemanticTermsMapper.class);
        job.setMapOutputKeyClass(Text.class);
        job.setMapOutputValueClass(IntWritable.class);
        job.setOutputKeyClass(Text.class);
        job.setOutputValueClass(IntWritable.class);
        job.setReducerClass(SemanticTermsReducer.class);
        return job.waitForCompletion(true);
}
// Sorting the top semantic terms.
private boolean cleanAndSorting(String input, String output)
              throws IOException, ClassNotFoundException, InterruptedException {
        Job job = Job.getInstance(getConf(), "cleanAndSorting");
        job.setJarByClass(this.getClass());
        FileInputFormat.setInputPaths(job, new Path(input));
        FileOutputFormat.setOutputPath(job, new Path(output));
        job.setOutputKeyClass(Text.class);
```

```
            job.setOutputValueClass(Text.class);
            job.setMapperClass(RankSortMapper.class);
            job.setMapOutputKeyClass(IntWritable.class);
            job.setMapOutputValueClass(Text.class);
            job.setOutputFormatClass(TextOutputFormat.class); // setting the output
                  // format of the
                  // clean and sorting
                  // job back to Text
                  // format.
            job.setSortComparatorClass(SortComparator.class); // Sort comparator
                  // class to sort the
                  // page rank results
                  // in the descending
                  // order.
            job.setNumReduceTasks(1); // setting the number of reduce tasks to be 1
            job.setReducerClass(RankSortReducer.class);
            return job.waitForCompletion(true);
    }
    public List<String> runLDA(String input) {
            // Begin by importing documents from text to feature sequences
            ArrayList<String> topicList = new ArrayList<String>();
            try {
                   ArrayList<Pipe> pipeList = new ArrayList<Pipe>();
                   // Pipes: lowercase, tokenize, remove stopwords, map to features
                   pipeList.add(new CharSequenceLowercase());
                   pipeList.add(new CharSequence2TokenSequence(Pattern.compile("\
                                 \p{L}[\\p{L}\\p{P}]+\\p{L}")));
                   pipeList.add(new TokenSequence2FeatureSequence());
                   InstanceList instances = new InstanceList(new
                                              SerialPipes(pipeList));
                   instances.addThruPipe(new CsvIterator(new StringReader(input),
                                 Pattern.compile("^(\\S*)[\\s,]*(\\S*)[\\s,]*(.*)$
                                         "), 3, 2, 1)); // data,
                                                       // label,
                                                       // name
                                                       // fields
                   // Create a model with 5 topics, alpha_t = 0.01, beta_w = 0.01
                   // Note that the first parameter is passed as the sum over topics,
                   // while
                   // the second is the parameter for a single dimension of the
                   // Dirichlet prior.
                   int numTopics = 5;
                   ParallelTopicModel model = new ParallelTopicModel(numTopics, 1.0,
                                              0.01);
                   model.addInstances(instances);

                   // Use two parallel samplers, which each look at one half the corpus
                   // and combine
                   // statistics after every iteration.
                   model.setNumThreads(2);
                   // Run the model for 50 iterations and stop (this is for testing
                   // only,
                   // for real applications, use 1000 to 2000 iterations)
                   model.setNumIterations(50);
                   model.estimate();
                   // Show the words and topics in the first instance
                   // The data alphabet maps word IDs to strings
                   Alphabet dataAlphabet = instances.getDataAlphabet();
                   FeatureSequence tokens = (FeatureSequence) model.getData().get(
                                 0).instance.getData();
                   LabelSequence topics = model.getData().get(0).topicSequence;
                   Formatter out = new Formatter(new StringBuilder(), Locale.US);
                   for (int position = 0; position < tokens.getLength(); position++)
{
                   out.format("%s-%d ", dataAlphabet.lookupObject(tokens.getIndexAtPo
                          sition(position)),
                                         topics.getIndexAtPosition(position));
                   }
                   // Estimate the topic distribution of the first instance,
                   // given the current Gibbs state.
                   double[] topicDistribution = model.getTopicProbabilities(0);
                   // Get an array of sorted sets of word ID/count pairs
                   ArrayList<TreeSet<IDSorter>> topicSortedWords = model.
                                                getSortedWords();
```

```
            // Show top 5 words in topics with proportions for the first
            // document
            for (int topic = 0; topic < numTopics; topic++) {
                    Iterator<IDSorter> iterator = topicSortedWords.g
                                        et(topic).iterator();
                    out = new Formatter(new StringBuilder(), Locale.US);
                    out.format("%d\t%.3f\t", topic, topicDistribution[topic]);
                    int rank = 0;
                    while (iterator.hasNext() && rank < 5) {
                            IDSorter idCountPair = iterator.next();
                            out.format("%s (%.0f) ", dataAlphabet.lookupO
                                bject(idCountPair.getID()), idCountPair.
                                getWeight());
                            rank++;
                            topicList.add((String) dataAlphabet.lookupObject
                                 (idCountPair.getID()));
            }
            }
            // Create a new instance with high probability of topic 0
            StringBuilder topicZeroText = new StringBuilder();
            Iterator<IDSorter> iterator = topicSortedWords.get(0).iterator();
            int rank = 0;
            while (iterator.hasNext() && rank < 5) {
                    IDSorter idCountPair = iterator.next();
      topicZeroText.append(dataAlphabet.lookupObject(idCountPair.getID()) + " ");
                            rank++;
                    }
                    // Create a new instance named "test instance" with empty
                       target and
                    // source fields.
                    InstanceList testing = new InstanceList(instances.
                                 getPipe());
                    testing.addThruPipe(new Instance(topicZeroText.
                            toString(), null, "test instance", null));
                    TopicInferencer inferencer = model.getInferencer();
                    double[] testProbabilities = inferencer.getSampledDist
                             ribution(testing.get(0), 10, 1, 5);
      } catch (Exception e) {
            e.printStackTrace();
      }
      return topicList;
}}
```

5 Optimization Approaches for Text Summarization

5.1 INTRODUCTION

Optimization is an important concept in making decisions and in analyzing automatic systems. In mathematical terms, an optimization problem is the problem of finding the best solution from among the set of all feasible solutions. Natural language processing (NLP) is influencing a rapid acceptance of more intelligent solutions in various end-use applications based on machine learning. Machine learning in turn has close connections to optimization: many learning problems are formulated as minimization of some loss functions such as error rate and noise reduction on a training set of examples. In this way, it is related to natural language or text analytics or processing applications such as text mining, text summarization, etc.

5.2 OPTIMIZATION FOR SUMMARIZATION

Optimization has been broadly used in the area of natural language processing and text analytics. There are many types of optimization problems in general. They can be classified as:

Continuous Optimization and Discrete Optimization

Some models only make sense if the variables take on values from a discrete set, often a subset of integers, whereas other models contain variables that can take on any real value. Models with discrete variables, the variables which could take finite, distinct, or separate values are discrete optimization problems; models with continuous variables, the variables which could take infinite, continuous values within a range are continuous optimization problems. Continuous optimization problems tend to be easier to solve than discrete optimization problems.

Unconstrained Optimization and Constrained Optimization

Another important distinction is between problems in which there are no constraints on the variables and problems in which there are constraints on the variables. Unconstrained optimization problems arise directly in many practical applications; they also arise in the reformulation of constrained optimization problems in which the constraints are replaced by a penalty term in the objective function. Constrained

DOI: 10.1201/9781003371199-5

optimization problems arise from applications in which there are explicit constraints on the variables.

Optimization Problems with Variable Number of Objectives

Most optimization problems have a single objective function; however, there are interesting cases when optimization problems have no objective function or multiple objective functions. Some of them are feasibility and complementarity problems. Feasibility problems are problems in which the goal is to find values for the variables that satisfy the constraints of a model with no particular objective to optimize. Complementarity problems are problems in which the goal is to find a solution that satisfies the complementarity conditions. Multi-objective optimization problems arise in many fields, such as engineering, economics, and logistics, when optimal decisions need to be taken in the presence of trade-offs between two or more conflicting objectives.

Deterministic Optimization and Stochastic Optimization

In deterministic optimization, it is assumed that the data for the given problem are known precisely. However, for many actual problems, the data cannot be known accurately for a variety of reasons. Stochastic programing models take advantage of the fact that probability distributions governing the data are known or can be estimated.

Any of the above models can be used to solve text summarization problems. They are detailed in the next section.

5.2.1 Modeling Text Summarization as Optimization Problem

Text summarization is commonly demonstrated as continuous or discrete optimization problem. They can also be viewed as single- or multi-objective non-constrained optimization problems.

The process includes defining two or more objectives based on some text features like redundancy, coverage, and relevance.

5.2.2 Various Approaches for Optimization

Many techniques using optimization have been employed for solving text summarization problems. They broadly fit into sentence ranking, evolutionary approach, MapReduce-based approach, and multi-objective approaches. Some screenshots of extractive and particle swarm optimization (PSO) summarization are shown in Figures 5.1 and 5.2.

5.3 FORMULATION OF VARIOUS APPROACHES

5.3.1 Sentence Ranking Approach

In the sentence ranking approach, optimization has to be done with the objective of ranking or selection of sentences from a document D consisting of a sequence of sentences $(s_1, s_2, \ldots, s_n)$.

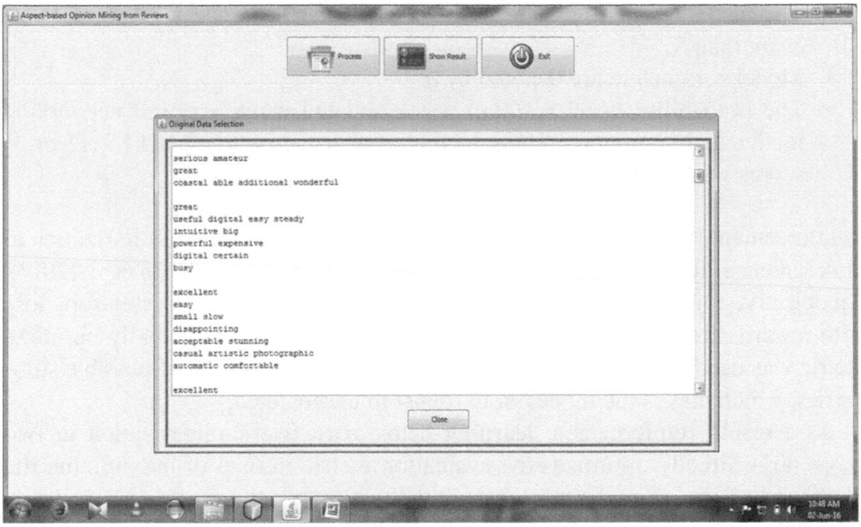

FIGURE 5.1 Extractive summarization

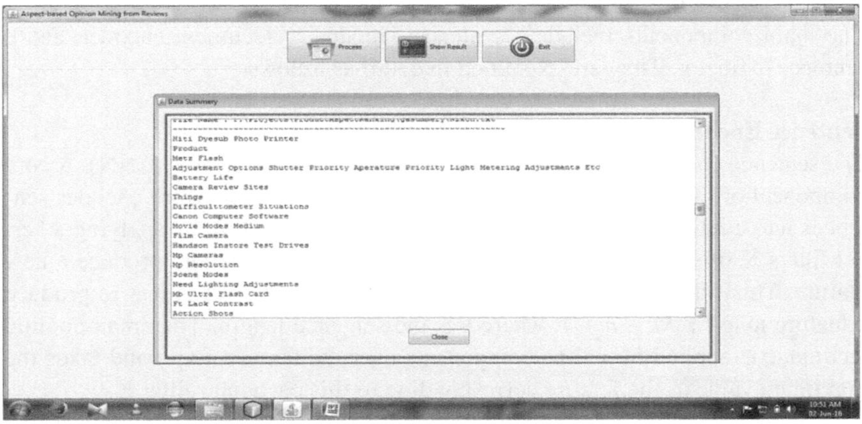

FIGURE 5.2 PSO summarization

The steps used for summarization based on the sentence ranking approach are as follows:

- An extractive summarizer aims to produce a summary S by selecting m sentences from D (where $m < n$).
- For each sentence, $S_i \in D$, a label $y_i \in \{0, 1\}$ (where 1 means that S_i should be included in the summary) was predicted.
- A score was assigned $p(y_i \mid S_i, D, q)$, quantifying relevance to the summary.

- The model learns to assign $p(1 \mid S_i, D, q)$ when sentence S_i is more relevant than S_j.
- Model parameters are denoted by q.
- The probability $p(y_i \mid S_i, D, q)$ was estimated using a neural network model, and a summary S by selecting m sentences with top $p(1 \mid S_i, D, q)$ scores was generated.

Reinforcement learning was adapted for formulating extractive summarization to rank sentences for generating summaries, which was given by Shashi et al. (2017). An objective function that combines the maximum-likelihood cross-entropy loss with rewards from policy gradient reinforcement learning to globally optimize metric was used. The training algorithm allows the exploration of possible summaries, which makes the model more robust to unseen data.

As a result, reinforcement learning helps extractive summarization in two ways: (a) it directly optimizes the evaluation metric instead of maximizing the likelihood of the ground-truth labels and (b) it makes the model better at discriminating among sentences; a sentence is ranked high for selection if it often occurs in high-scoring summaries. The overall framework is shown in Figure 5.3.

5.3.1.1 Stages and Illustration

The main components include a sentence encoder, a document encoder, and a sentence extractor. They are explained in detail as follows:

Sentence Encoder:

The sentence encoder is used with convolutional neural networks (CNN). A core component of the model is a convolutional sentence encoder which encodes sentences into continuous representations. A narrow convolution by applying a kernel filters K of width h to a window of h words in sentence s to produce a new feature. This filter is applied to each possible window of words in s to produce a feature map f Ɛ $Rk - h + 1$, where k is the sentence length. Then max-pooling technique is applied over time components over the feature map f and takes the maximum value as the feature corresponding to this particular filter K.

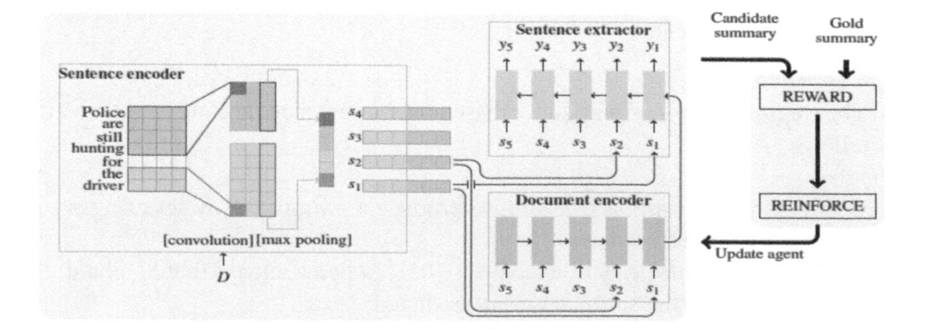

FIGURE 5.3 Reinforcement learning framework for summarization

Document Encoder:

The document encoder composes a sequence of sentences to obtain a document representation. A recurrent neural network is used with long short-term memory (LSTM) cells to avoid the vanishing gradient problem when training long sequences.

Sentence Extractor:

Sentence extractor sequentially labels each sentence in a document with 1 (relevant for the summary) or 0 (otherwise). It is implemented with another recurrent neural network (RNN) with LSTM cells and a softmax layer. At time t_i, it reads sentence S_i and makes a binary prediction, conditioned on the document representation (obtained from the document encoder) and the previously marked sentences with labels. This way, the sentence extractor is able to identify locally and globally important sentences within the document. The sentences are ranked in a document D by $p(y_i = 1 \mid S_i, D, q)$, the confidence scores assigned by the softmax layer of the sentence.

These are the three main stages in the framework. As a whole, the model can be viewed as an 'agent' who interacts with an 'environment' consisting of documents. At first, the agent is initialized randomly, it reads document D and predicts a relevance score for each sentence S_i using 'policy' $p(y_i \mid S_i, D, q)$, where q are model parameters. Once the agent is done reading the document, a summary with label ^ y is sampled out of the ranked sentences. The agent is then given a 'reward' r commensurate with how well the extract resembles the gold-standard summary.

Specifically, as reward function, mean F_1 of the metrics like Recall-Oriented Understudy for Gisting Evaluation (ROUGE)-1, ROUGE-2, and ROUGE-L was used. The model generates an informative, fluent, and concise summary outperforming state-of-the-art extractive systems on the CNN and Daily Mail datasets.

5.3.2 Evolutionary Approaches

Artificial intelligence–based techniques such as particle swarm optimization, genetic algorithm, artificial bee colony optimization, and ant colony optimization are population-based heuristic search methods used for solving large-scale problems. These methods are found to improve the relevancy score and reduce the computational cost in text summarization problems. These algorithms can be applied in both single-document and multi-document summarization problems. Among these algorithms, cuckoo search () was one of the recent methods and was successful for all cases.

5.3.2.1 Stages

Ali et al. (2019) gave an optimization approach for text summarization based on the cuckoo search algorithm. In the first stage, significant features are extracted for each sentence in the document collection. In the next stage, the cuckoo search algorithm was applied to compute weights for these extracted features. Finally, the algorithm based on multi-criteria decision-making, named VIKOR, is used to rank the sentences. VIKOR is expanded as 'VlseKriterijumska Optimizacija I Kompromisno Resenje' meaning multi-criteria optimization and compromise solution is an algorithm used for ranking. The sentences with a high score and less redundancy are selected to be included in the final summary. All the stages are explained in the following section.

5.3.2.2 Demonstration

The block diagram of the model for VIKOR algorithm–based cuckoo search for summarization is shown in Figure 5.4. All the stages are detailed as follows:

Feature Extraction:

After preprocessing steps such as sentence segmentation, tokenization, stop word removal, and stemming, as a next step the features are extracted from the documents. The features considered are as follows:

i. Sentence position (SP): Higher score will be given to the first sentence; the score decreases according to the sentence position in the document. This feature can be computed according to Equation (5.1).

$$F_1(S_i) = \frac{N - i + 1}{N} \tag{5.1}$$

where i is the position of the sentence (S) in a document of N sentences

ii. Sentence length (SL): This feature is computed by dividing the sentence length by the length of the longest sentence in the document as in Equation (5.2).

$$F_2(S_i) = \frac{L(S_i)}{L_{\max}} \tag{5.2}$$

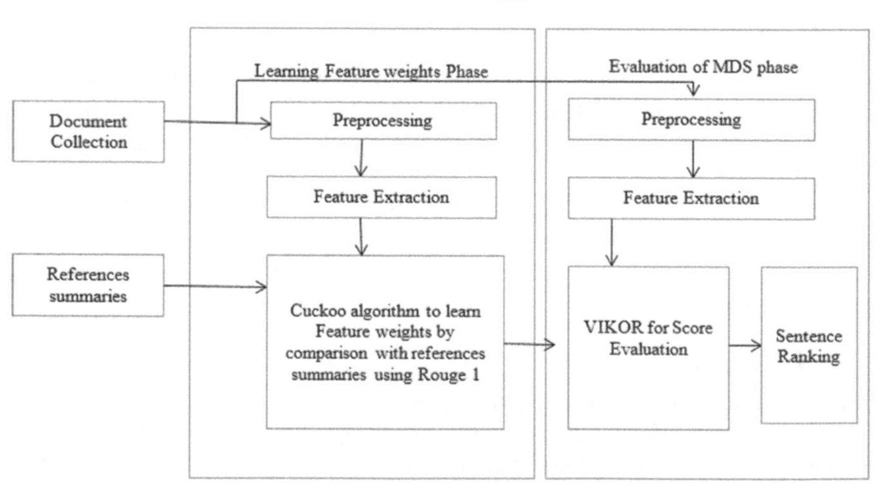

FIGURE 5.4 VIKOR-based cuckoo search model for summarization

where $L(S_i)$ is the length of the sentence and L_{max} is the length of the longest sentence in the document.

iii. Numerical data (ND): It has important information to be included in the summary. This feature is calculated by dividing the number of numerical data in the sentence divided by the length of the sentence as in Equation (5.3).

$$F_3(S_i) = \frac{Num(S_i)}{L(S_i)} \tag{5.3}$$

where Num(S_i) is the number of numerical data in the sentence (S_i)

iv. Thematic words (TW): These are the terms that appear more frequently than other terms in the document. TW can be calculated as in Equation (5.4).

$$F_4(S_i) = \frac{\text{TW}(S_i)}{\text{TW}_{max}} \tag{5.4}$$

where TW is the number of thematic words in the sentence and TW_{max} is the maximum number of thematic words in the sentences.

v. Title feature (TF): This feature is computed by finding the overlap between the title sentence and every sentence in the document as in Equation (5.5).

$$F_5(S_i) = \frac{\text{No of TF}}{L(\text{Title})} \tag{5.5}$$

where TF is the words that exist in both S_i and Title

vi. Proper noun (PN): It is based on computing the number of PN in every sentence as in Equation (5.6).

$$F_5(S_i) = \frac{\text{PN}(S_i)}{L(S_i)} \tag{5.6}$$

where PN is the number of proper nouns in a sentence S_i.

The above six features for each solution can be represented as six bits, each bit corresponding to one feature value. The bit value can be either 1 or 0 – 1 means the corresponding features will be selected and 0 means the corresponding features will be ignored.

VIKOR Scoring Phase

VIKOR technique can be considered as a compromise ranking method; it is based on ranking and choosing a suitable alternative from a set of alternatives. Each alternative is evaluated according to each criterion function, then choose the alternative that is close to the ideal solution. The features extracted in the feature extraction phase are considered alternatives in the VIKOR decision matrix. The goal of the VIKOR algorithm is to rank all sentences based on these features from the best to worst one. The best sentences are sentences that include a high combination of features. All these methods are based on scoring the sentences, and the sentences with higher scores are chosen to be included in the final summary. According to this hypothesis, some features may have more effect than other features; therefore, sentences with higher values of these features have more chance to be included in the final summary. At the end of the training and scoring procedure, a set of good solutions is created. The average of these solutions is computed by summing every bit with its corresponding bits in other solutions, then dividing them by the total number of solutions. Finally summary sentences are selected.

There are many documents with the same topic, but some sentences may be repeated in more than one document. A technique is required to remove the redundant sentences from the generated summary, which allows the final summary to include the most important ideas for the summarized documents. The cosine similarity is used to compute the similarity between two sentences and exclude the sentence from a final summary when the similarity between them is more than a specified threshold. The results were better than the results of the other summaries because of the performance of VIKOR algorithm and the choice of features.

5.3.3 MapReduce-Based Approach

In recent times, several optimization techniques (2017) utilizing MapReduce framework have been studied successfully by researchers. They are query optimization, performance optimization, application-level optimization, smart grids optimization, parameter optimization, and so on. Algorithm-level optimization works on combiner-based optimization approach. The two key problems with combiner modules in MapReduce [8] are:

- Execution of a combiner is not guaranteed: Combiners may not be executed on some occasions as Hadoop may choose not to run them if execution is determined to be inefficient for the system.
- Size of emitted map outputs is not optimized: The emitted results are temporarily stored in in-memory buffers, and the combining function is applied to them before spilling them to the local disk. Thus combiners do not actually reduce the number of emitted results. This characteristic leads to situations where map output buffers are filled with soon-to-be combined outputs causing more spill files to be generated.

The in-mapper combiner (IMC) resolves the two problems of the traditional combiners. The key idea behind IMC is to run the combining function inside the map method. This minimizes the volume of emitted intermediate results. Instead of emitting results to map output buffers at every invocation of the MAP method,

IMC stores and aggregates results in an associative array indexed by output keys. They emit them at the end of the map task. This approach guarantees the execution of combiners and substantial reduction in the total number of emitted map outputs. Priya et al. (2019a) [6] utilized this for optimized aspect-based text summary generation. This has been discussed in the next section.

5.3.3.1 In-Node Optimization Illustration

In this method Hadoop stores the input file as sequence file comprising <key, value> pairs. The key is the text or aspect in the file, and the value string contains related reviews for one particular aspect along with the relevance score based on tf-idf model.

The main function of in-node mapper is to cluster the mapper output pertaining to identical key. A combiner class is used along with map class as in-node combiner for optimization purposes. The process in the mapper function is as follows: All the review sentences along with the key terms are stored in hash map. In cleanup method, the sentences which are more similar to the first representative sentence are not spilled in the disk. Similarity is computed using the metric given by Wael et al. [9]. This minimizes the data transfer among the nodes, and also redundant sentences are eliminated. This ensures that redundant sentences are not included which improves the quality of the summary. The detailed algorithm for in-node mapper is shown as follows:

```
Algorithm: InnodeMapper (String array, a[i], String array,
list of sentences)
Input: key [text (aspect keywords, a[i])], value [review
sentences for one aspect]
Output: <key', value'> pair, where the key' is the feature
or aspect term, Ak and value' - filtered output review
sentences based on relevance threshold.
Class Mapper {
Method map(Text, Text) {
Map < String, String> AspectMap = new Hash Map <String,
String> ( );
For all keys do
{ AspectMap.put(key_id,Value); }}
method Cleanup ()
{       for Map.Entry <String, String>
        entry: featureMap.entrySet() {
                 if (sim(sid1,value)) < 0.5 then emit <
                    key', value' > pair
                 else discard values}
}}
```

The output from the in-node combiner with mapper is partitioned using a custom partitioner to segregate the summaries for all the significant aspects identified. The number of reducer tasks is dependent on the important aspects identified from different domains. The reducer iterates through each key and writes with the corresponding aspect key term. The algorithm is run for three various domains. The algorithm is given as follows:

```
public class aspectPartitioner extends MapoutputPartitioner
<Text, Text>{
public int getmapoutputPartition( Text key, Text value, int
numReduceTasks)
{
If Text = key1 then
return 1 % numReduceTasks;
elseif Text = key2 then
return 2 % numReduceTasks
elseif Text = key3 then
return 3 % numReduceTasks
.... // continue for all the aspect keys
}}
Static Class PartitionReducer extends InmapperReducer (Text,
Text, Text, Text)
{
public void sumreduce (Text key', iterable <Text> value'){
For (Text val:values){
String [ ] valTokens = val.toString().split(",");
Context.write (new (Text (key'), new (Text (value'))) );
}}
```

This method when evaluated with reviews from the movie domain and hotel domain achieved a better result than other modern methods. The next section explains about multi-objective-based summarization method.

5.3.4 Multi-objective-Based Approach

There had been various multi-objective evolutionary methods found for solving text summarization problem. The method given by Naveen et al. (2018) was detailed in this section along with other variations recently found in the literature.

Methods which try to solve two or more objectives could be categorized as multi-objective methods. In Naveen's approach, multi-objective optimization (MOO) framework had been used for text summarization problem. Three different techniques found in their research are (i) integration of self-organizing map and multi-objective differential evolution, (ii) multi-objective gray wolf optimizer, and (iii) multi-objective water cycle algorithm. The sentences present in the document are first clustered utilizing the concept of multi-objective clustering. Two objective functions measuring compactness and separation of the sentence clusters in two different ways are optimized simultaneously using MOO framework.

i. **Multi-objective optimization (MOO)-based clustering**

Sentence cluster can be encoded using a representative sentence in the solution. A solution xi encodes a set of different cluster centers. As it is difficult to decide the number of clusters present in the document, the number of cluster centers in each solution is varied over a range, $1 \leq K \leq N$, where K is the number of clusters and N is the number of sentences in the document. To generate the ith solution, a random number is generated between $1 \leq K \leq N$ and then K number of sentences are selected as initial cluster centers. In order to extract sentence clustering corresponding to the ith solution, K-medoid algorithm is used with the present set of cluster centers. After each iteration of K-medoid algorithm, cluster representatives/centroids are updated, and this process continues until cluster centroids converge. Note that in K-medoid, the cluster representatives/centroids will be the sentences belonging to the document. Thus population P consists of solutions $\langle x_1, x_2, \ldots, x|P|\rangle$ with varying numbers of clusters. As solutions in the population further take part in Self Organizing Maps (SOM) training, variable length solutions are converted into fixed length by appending zeros. After forming the population, solutions (solution space, not objective space) in the population take part in SOM training. Three types of genetic operators used are mating pool selection, crossover/repairing, and mutation. They are applied to generate new solutions from the present set of solutions on the population.

After generating the new solutions using the genetic operators, a new population P' is formed which is further merged with the old population P. Thus total solutions in the merged population will be $2 \times |P|$, where $|P|$ is the size of the population. In the next generation, only the best $|P|$ solutions are passed which are selected using the domination and non-domination relationships in the objective function space of the solutions. Arrange sentences in descending order present in a cluster according to their sentence scores. Now, to generate summary, clusters are considered rank-wise. Given a cluster, top-ranked sentences are extracted sequentially until summary length reaches to some threshold.

ii. **Multi-objective gray wolf optimizer**

Gray wolf algorithm is based on leadership hierarchy and hunting procedure of gray wolves in nature. Both the leader hierarchy and hunting procedure are detailed as follows:

Leadership hierarchy:

i. Alpha is the leader of the group and is mostly responsible for making decisions about sleeping, hunting, and so on.
ii. Beta wolves help the alpha in decision-making and other activities of the pack/group. They also maintain discipline in the pack.
iii. Delta wolves obey alpha and beta wolves. They are generally elders, sentinels, and caretakers in the pack.
iv. Omega wolves are the lowest-ranked wolves, and they respect alpha, beta, and delta wolves.
v. Note that alpha is not the strongest member of the pack, but the best in terms of decision-making and managing the pack.

vi. Beta wolves have the potentiality to become alpha in case alpha passed away or became very old.
vii. In terms of the evolutionary algorithm, alpha (α) is the best solution in the population.
viii. The second and third best solutions are termed as beta (β) and delta (δ), respectively.
ix. Rest of the solutions are called omega (ω) wolves.

Hunting technique of wolves:

The hunting techniques of wolves include (a) chasing and approaching the prey, (b) harassing and encircling the prey until it stops, and (c) attacking the prey. Note that in this case, prey is the most optimal solution in terms of objective functional value. The wolves update their positions and hunt towards the prey based on objective functions till convergence.

iii. **Multi-objective water cycle algorithm**

Each solution is associated with objective functional values or fitness values. The fittest solution is considered as the sea (represented as solutions).

The Nsr solutions are considered as rivers and remain as streams, where Nsr is the summation of the number of rivers (which is defined by the user) and a single sea.

Here Nsr = 1 sea + Number_of _rivers, where NSn is the number of streams that flow into a specific river and sea.

To show the flow of streams to rivers, Equation (5.7) is considered:

$$\vec{x}stream(t+1) = \vec{x}stream(t) + R \times C \times (\vec{x}river(t) - \vec{x}stream(t)) \tag{5.7}$$

where $1 < C < 2$ and R lies between [0, 1], $\vec{x}stream(t+1)$ represents the updated position of the stream and $\vec{x}stream$ at time $(t + 1)$, $\vec{x}river(t)$ shows the position of river at time t.

Similarly, Equations 5.8 and 5.9 can be written to update the position of the river in case the river flows to the sea.

$$\vec{x}river(t+1) = \vec{x}river(t) + R \times C \times (\vec{x}sea(t) - \vec{x}river(t)) \tag{5.8}$$

and update the position of the stream in case the stream flows to the sea.

$$\vec{x}stream(t+1) = \vec{x}stream(t) + R \times C \times (\vec{x}sea(t) - \vec{x}stream(t)) \tag{5.9}$$

If solution given by stream (after updating position) is better than its connecting river, then stream and river exchange their positions.

Similar steps can be executed between stream and sea, and river and sea.

After updating the position, evaporation condition is checked to generate new solutions, i.e. whether streams/rivers are close to sea within a radius to make the evaporation process to occur as given per Equation 5.10 must be verified:

$$\|\vec{x}sea - \vec{x}river\| < dmax \text{ or } rand() < 0.1 \quad (5.10)$$

where dmax is a small number close to zero and linearly decreases over the course of iteration.

Here, dmax is responsible for exploitation near the sea as the sea is considered an optimal solution.

After evaporation, new streams are formed at different locations due to raining process.

This step is like exploration in search space.

The new stream generated can be shown in Equation (5.11) as:

$$\vec{\mathbf{x}}\mathbf{steam\ new} = \vec{\mathbf{l}}\mathbf{b} + r_1 \times (\vec{\mathbf{u}}\mathbf{p} - \vec{\mathbf{l}}\mathbf{b}) \quad \mathbf{(5.11)}$$

where r_1 is the random number between [0, 1], $\vec{l}b$ and $\vec{u}b$ are the lower and upper bounds given by the problem.

Water cycle algorithm is a meta-heuristic algorithm that mimics the water cycle process in nature, i.e. the flow of rivers and streams to sea and flow of streams to rivers. It is a nature-inspired population-based algorithm. Therefore, it needs to generate an initial population containing a set of solutions. The steps in the algorithm are given below.

Thus, these steps are executed over a fixed number of iterations to search for the optimal solution, i.e. the sea. This process of the water cycle algorithm was executed by Sadollah et al. (2015). Their system was evaluated using DUC 2001 and DUC 2002 datasets. Results showed the efficacy of the three evolutionary or nature-inspired algorithms when compared to other evolutionary algorithms.

There are other types of optimization methods in the literature which is worth when the mentioned are constrained optimization which was given by Rasim et al. (2015).

```
Initialize parameters of MPSOAS
Randomly initialize swarm and partition into two sets
While stopping criterion not met do
        Calculate each particle's fitness function
                for i=1 to swarmSize do
                        Update the local best of pi
                        Update the previous best of pi
                end
```

```
                    for i = 1 to swarmSize do
                    for j = 1 to dimension do
                              update the velocity of pi
                                according to Equation (5.12)
                              update the position of pi
                                according to Equation (5.13)
                    end
                    end
                    If diversity < d min, then
                    Keep some particles' (α percent) position
                         and velocity, re-initialize others
                    randomly after each β iteration.
                    Keep some particles' (β percent) position
                         and velocity, re-initialize others
                    randomly after each γ iteration.
                    Keep some particles' (γ percent) position
                         and velocity, re-initialize others
                    randomly after each α iteration.
                              Again update the velocity of pi
                               according to Equation (5.12)
                              Again update the position of pi
                               according to Equation (5.13)
                    Else
                              Compute fitness of all the
                               particles using Equation (5.14)
End while
Return the best particle subset found by the swarm from the
       archive.
```

Continuous and discrete optimization methods using particle swarm optimization (PSO) were given by many authors such as Giannakopoulos et al., 2011; Jena et al., 2017; Lee et al., 2015; Wael et al., 2013; Sanchez et al., 2018; Jain and Dubes 1988; Eskandar et al., 2012; Priya and Umamaheswari 2019b. In this method, the particle swarm optimization algorithm using improved diversity had been developed and tested with these two representations. Continuous and discrete models were established and verified with customer reviews. The algorithm used is shown below.

The equations for computing velocity, position, and fitness are as given by Equations 5.12–5.14.

$$(i) = \text{Max}(P_i(\text{tf-idf}) + \text{SO}(P_i)) \tag{5.12}$$

where P_i(tf-idf) = Term frequency of the ith particle
SO(P_i) = Opinion score of the ith particle

$$X_i = ((V_i(t_{\max}), P_i)) \tag{5.13}$$

Here similarity between the sentence with maximum term frequency and the new sentence as a particle is calculated. If the similarity is found minimum, then the position will be updated, since it is a new sentence.

$$\max\left(F\left(x\right)\right)=\sum_{j=1}^{n}\begin{matrix}\max\left(w_{ij}\right)\text{where } 1< w_{ij}<10\\ \min(\text{sim}\left(p_{ij}\right)\text{where }\left(\text{sim}\left(p_{ij}\right)\right)<10\end{matrix} \tag{5.14}$$

The weights w_{ij} are assigned as 1–10 for top n features identified previously. The maximum tf-idf score is taken as reference from the velocity of the particle. The weight score is normalized using probability values on a scale of 0.9–0.1 which indicates maximum and minimum, respectively. The component sim(p_{ij}) denotes the similarity of ith particle in jth dimension with maximum term score for ith sentence.

Real-valued continuous model with PSO had better performance in the quality of the summary. This is measured with ROUGE metrics and increased by 5% compared to other existing systems.

The method explained in Section 5.3.3.1 for summarization which involved in-node combiner base optimization had also shown improved results. In this method, the noisy sentences were removed from the reviews using optimization, and also this minimized processing time substantially. The quality of the summary also displays an improvement around 6% when using ROUGE metrics.

Cognitive-based optimization:
Elayeb et al. (2020) proposed a new extractive single-document summarization approach based on analogical proportions which are statements of the form 'a is to b as c is to d.' The goal is to study the capability of analogical proportions to represent the relationship between documents and their corresponding summaries. For this purpose, they suggested two algorithms to quantify the relevance/irrelevance of an extracted keyword from the input text, to build its summary. In the first algorithm, the analogical proportion representing this relationship is limited to check the existence/non-existence of the keyword in any document or summary in a binary way without considering keyword frequency in the text, whereas the analogical proportion of the second algorithm considers this frequency. This model can be used for cognition-based summarization for optimizing the sentences and summary lengths.

SUMMARY

Optimization algorithms tend to produce efficient results for text summarization in the literature. Though this is proven, there are some factors that could be improved when applied to text summarization problem. They are sentence ordering, parameter tuning, convergence criteria, automatic weight adaptation from the algorithm, and so on. This extends the applications of optimization algorithms to various directions for solving text summarization problem.

EXERCISES

Example 1:

1. Find the maximum of the function $f(x) = -x^2 + 5x + 20$ with $-10 \leq x \leq 10$ using the PSO algorithm. Use nine particles with the initial positions $x_1 = -9.6$, $x_2 = -6$, $x_3 = -2.6$, $x_4 = -1.1$, $x_5 = 0.6$, $x_6 = 2.3$, $x_7 = 2.8$, $x_8 = 8.3$, and $x_9 = 10$. Show the detailed computations for iterations 1, 2, and 3.

Solution:

Step 1: Choose the number of particles:

$x_1 = -9.6$, $x_2 = -6$, $x_3 = -2.6$, $x_4 = -1.1$, $x_5 = 0.6$, $x_6 = 2.3$, $x_7 = 2.8$, $x_8 = 8.3$, and $x_9 = 10$

The initial population (i.e. the iteration number) can be represented as:

x_i^0, $i = 1, 2, 3, 4, 5, 6, 7, 8, 9$

$x_1^0 = -9.6$, $x_2^0 = -6$, $x_3^0 = -2.6$, $x_4^0 = -1.1$, $x_5^0 = 0.6$, $x_6^0 = 2.3$, $x_7^0 = 2.8$, $x_8^0 = 8.3$, and $x_9^0 = 10$

Evaluate the objective function values as

$f_1^0 = -120.16$, $f_2^0 = -46$, $f_3^0 = 0.24$, $f_4^0 = 13.29$, $f_5^0 = 22.64$, $f_6^0 = 26.21$, $f_7^0 = 26.16$, $f_8^0 = -7.39$, and $f_9^0 = -30$

Let $c_1 = c_2 = 1$. Set the initial velocities of each particle to zero $v_i 0 = 0$, i.e. $v_1^0 = v_2^0 = v_3^0 = v_4^0 = v_5^0 = v_6^0 = v_7^0 = v_8^0 = v_9^0 = 0$.

Step 2: Set the iteration number as $t = 0 + 1 = 1$ and go to step 3.

Step 3: Find the personal best for each particle by:

$$P_{\text{best},i}^{t+1} = \begin{cases} P_{\text{best},i}^{t} & \text{if } f_i^{t+1} > P_{\text{best},i}^{t} \\ x_i^{t+1} & \text{if } f_i^{t+1} > P_{\text{best},i}^{t} \end{cases}$$

So,

$P_{\text{best},1}^1 = -9.6$, $P_{\text{best},2}^1 = -6$, $P_{\text{best},3}^1 = -2.6$, $P_{\text{best},4}^1 = -1.1$, $P_{\text{best},5}^1 = 0.6$, $P_{\text{best},6}^1 = 2.3$, $P_{\text{best},7}^1 = 2.8$, $P_{\text{best},8}^1 = 8.3$, $P_{\text{best},9}^1 = 10$

Step 4: Find the global best by

$G_{best} = \min\left\{P_{\text{best},i}^t\right\}$ where $i = 1, 2, 3, 4, 5, 6, 7, 8, 9$.

Since the maximum personal best is $P_{\text{best},6}^1 = 2.3$, thus $G_{\text{best}} = 2.3$.

Step 5: Considering the random numbers in the range (0, 1) as $r_1^1 = 0.213$ and $r_1^1 = 0.876$, find the velocities of the particles by

$$v_{t+1}^{i} = v_{t}^{i} + c_1 r_1^t \left[P_{\text{best},i}^{t} - x_i^t \right] + c_2 r_2^t \left[G_{\text{best},i}^{t} - x_i^t \right]; i = 1,2\ldots,9$$

So,

$v_1^1 = 0 + 0.213(-9.6 + 9.6) + 0.876(2.3 + 9.6) = 10.4244$

$v_2^1 = 7.2708$, $v_3^1 = 4.2924$, $v_4^1 = 2.9784$, $v_5^1 = 1.4892$, $v_6^1 = 0$, $v_7^1 = -0.4380$, $v_8^1 = 5.256$, and $v_9^1 = -6.742$

Step 6: Find the new values of x_i^1, $i = 1,2,3,\ldots 9$ by

$x_i^{t+1} = x_i^t + v_{t+1}^i$ so, $x_1^1 = 0.8244$, $x_2^0 = 1.2708$, $x_3^0 = 1.6924$, $x_4^0 = 1.8784$, $x_5^0 = 2.0892$, $x_6^0 = 2.3$, $x_7^0 = 2.362$, $x_8^0 = 3.044$, and $x_9^0 = 3.2548$

Step 7: Find the objective function values of x_i^1, $i = 1, 2, 3, 4, 5, 6, 7, 8, 9$:

$f_1^1 = 23.4424$, $f_2^1 = 24.7391$, $f_3^1 = 25.5978$, $f_4^1 = 25.8636$, $f_5^1 = 26.0812$,

$f_6^1 = 26.21$, $f_7^1 = 26.231$, $f_8^1 = 25.9541$, and $f_9^1 = 25.6803$

Step 8: Stopping criterion:

If the terminal rule is satisfied, go to step 2,

Otherwise stop the iteration and output the results.

Step 2: Set the iteration number as $t = 1 + 1 = 2$ and go to step 3.

Step 3: Find the personal best for each particle by:

$P_{\text{best},1}^{2} = 0.8244$, $P_{\text{best},2}^{2} = 1.2708$, $P_{\text{best},3}^{2} = 1.6924$, $P_{\text{best},4}^{2} = 1.8784$, $P_{\text{best},5}^{2} = 2.0892$, $P_{\text{best},6}^{2} = 2.3$, $P_{\text{best},7}^{2} = 2.62$, $P_{\text{best},8}^{2} = 3.044$, $P_{\text{best},9}^{2} = 3.2548$

Step 4: Find the global best $G_{\text{best}} = 2.362$

Step 5: Considering the random numbers in the range (0, 1) as $r_1^1 = 0.113$ and $r_1^1 = 0.706$, find the velocities of the particles by

$$v_{t+1}^{i} = v_{t}^{i} + c_1 r_1^t \left[P_{\text{best},i}^{t} - x_i^t \right] + c_2 r_2^t \left[G_{\text{best},i}^{t} - x_i^t \right]; i = 1,2\ldots,9$$

So,

$v_1^2 = 11.5099$, $v_2^2 = 8.0412$, $v_3^2 = 4.7651$, $v_4^2 = 3.3198$, $v_5^2 = 1.68182$, $v_6^2 = 0.0438$, $v_7^2 = -0.4380$, $v_8^2 = -5.7375$, and $v_9^2 = -7.3755$

Step 6: Find the new values of x_i^1, $i = 1,2,3,\ldots 9$ by

$$x_i^{t+1} = x_i^t + v_{t+1}^i$$

So, $x_1^2 = 12.3343$, $x_2^2 = 9.312$, $x_3^2 = 6.4575$, $x_4^2 = 5.1982$, $x_5^2 = 3.7710$, $x_6^2 = 2.3438$, $x_7^2 = 1.9240$, $x_8^2 = -2.6935$, and $x_9^2 = -4.1207$

Step 7: Find the objective function values of f_i^2, $i = 1, 2, 3, 4, 5, 6, 7, 8, 9$:

$f_1^2 = -70.4644$, $f_2^2 = -20.1532$, $f_3^2 = 10.5882$, $f_4^2 = 18.9696$, $f_5^2 = 24.6346$,

$f_6^2 = 26.2256$, $f_7^2 = 25.9182$, $f_8^2 = -0.7224$, and $f_9^2 = -17.5839$

Step 8: Stopping criterion:

Step 2: Set the iteration number as $t = 2 + 1 = 3$ and go to step 3.
Step 3: Find the personal best for each particle

So,

$P^3_{\text{best},1} = 0.8244$, $P^3_{\text{best},2} = 1.2708$, $P^3_{\text{best},3} = 1.6924$, $P^3_{\text{best},4} = 1.8784$, $P^3_{\text{best},5} = 2.0892$, $P^3_{\text{best},6} = 2.3438$, $P^3_{\text{best},7} = 2.362$, $P^3_{\text{best},8} = 3.044$, $P^3_{\text{best},9} = 3.2548$

Step 4: Find the global best

$$G_{\text{best}} = 2.362$$

Step 5: By considering the random numbers in the range (0, 1) as $r_1^3 = 0.178$ and $r_1^3 = 0.507$, find the velocities of the particles by

$$v^i_{t+1} = v^i_t + c_1 r^t_1 \left[P^t_{\text{best},i} - x^t_i \right] + c_2 r^t_2 \left[G^t_{\text{best},i} - x^t_i \right]; i = 1,2\ldots,9$$

So,

$v_1^3 = 4.4052$, $v_2^3 = 3.0862$, $v_3^3 = 1.8405$, $v_4^3 = 1.2909$, $v_5^3 = 0.6681$, $v_6^3 = 0.053$, $v_7^3 = -0.1380$, $v_8^3 = -2.1531$, and $v_9^3 = -2.7759$

Step 6: Find the new values of x^1_i, $i = 1,2,3,\ldots 9$ by

$$x^{t+1}_i = x^t_i + v^i_{t+1}$$

So,

$x_1^3 = 16.7395$, $x_2^3 = 12.3982$, $x_3^3 = 8.298$, $x_4^3 = 6.4892$, $x_5^3 = 4.4391$, $x_6^3 = 2.3968$, $x_7^3 = 1.786$, $x_8^3 = -4.8466$, and $x_9^3 = -6.8967$

Step 7: Find the objective function values of f_i^3, $i = 1, 2, 3, 4, 5, 6,7, 8 , 9$:

$f_1^3 = -176.5145$, $f_2^3 = -71.7244$, $f_3^3 = -7.3673$, $f_4^3 = 10.3367$, $f_5^3 = 22.49$,

$f_6^3 = 26.2393$, $f_7^3 = 25.7402$, $f_8^3 = -27.7222$, and $f_9^3 = -62.0471$

Step 8: Stopping criterion:

Finally, if the values of did not converge, so we increment the iteration number as and go to step 2. When the positions of all particles converge to similar values, then the method has converged and the corresponding value of f is the optimum solution. Therefore, the iterative process is continued until all particles meet a single value.

Example 2:

Given the following input text, give the output for all the phases including combiner. Assume you are running word count program.

Input Text:

What do you mean by Object
What do you know about Java

What is Java Virtual Machine
How Java enabled High Performance

Record Reader Output:

<1, What do you mean by Object>
<2, What do you know about Java>
<3, What is Java Virtual Machine>
<4, How Java enabled High Performance>

Mapper Output:

<What,1> <do,1> <you,1> <mean,1> <by,1> <Object,1>
<What,1> <do,1> <you,1> <know,1> <about,1> <Java,1>
<What,1> <is,1> <Java,1> <Virtual,1> <Machine,1>
<How,1> <Java,1> <enabled,1> <High,1> <Performance,1>

Combiner Output:

<What,1,1,1> <do,1,1> <you,1,1> <mean,1> <by,1> <Object,1>
<know,1> <about,1> <Java,1,1,1>
<is,1> <Virtual,1> <Machine,1>
<How,1> <enabled,1> <High,1> <Performance,1>

Reducer Output:

<What,3> <do,2> <you,2> <mean,1> <by,1> <Object,1>
<know,1> <about,1> <Java,3>
<is,1> <Virtual,1> <Machine,1>
<How,1> <enabled,1> <High,1> <Performance,1>

Exercises:

1. Read the original MapReduce paper by Dean and Ghemawat [1]. You can find the paper at http://research.google.com/archive/mapreduce.html
   Answer the following questions:
   a) How do the input keys/values, the intermediate keys/values, and the output keys/values relate?
   b) How does MapReduce deal with node failures?
   c) What is the meaning and the implication of locality? How is it used?
   d) Which problem is addressed by introducing a combiner function to the MapReduce model?
2. Consider the input data given for MapReduce as follows:
   Car beer apple
   Apple beer beer Mango

Mango car car
Beer apple mango

For all the stages illustrate the outputs, assuming you are running word count program with combiner.

REFERENCES

Alguliyev, R. M., Aliguliyev, R. M., and Isazade, N. R. 2015. "An unsupervised approach to generating generic summaries of documents." *Applied Soft Computing* 34: 236–250.

Ali, Z. H., Noor, A. A., and Jassim, M. A. 2019. "VIKOR algorithm based on cuckoo search for multi-document text summarization." In *International Conference on Applied Computing to Support Industry: Innovation and Technology* (pp. 57–67). Cham: Springer.

Elayeb, B., Chouigui, A., Bounhas, M., and Khiroun, O. B. 2020. "Automatic Arabic text summarization using analogical proportions." *Cognitive Computation* 12(5): 1043–1069.

Eskandar, H., Sadollah, A., Bahreininejad, A., and Hamdi, M. 2012. "Water cycle algorithm–a novel metaheuristic optimization method for solving constrained engineering optimization problems." *Computers and Structures* 110: 151–166.

Giannakopoulos, G., El-Haj, M., Favre, B., Litvak, M., Steinberger, J., and Varma, V. 2011. *TAC 2011 MultiLing pilot overview.* Maryland, USA: TAC.

Gomaa, W. H., and Fahmy, A. A. 2013. "A survey of text similarity approaches." *International Journal of Computer Applications* 68(13): 13–18.

Jain, A. K., and Dubes, R. C. 1988. *Algorithms for Clustering Data.* Hoboken, New Jersey ,USA: Prentice-Hall, Inc.

Jena, B., Gourisaria, M. K., Rautaray, S. S., and Pandey, M. 2017. "A survey work on optimization techniques utilizing map reduce framework in hadoop cluster." *International Journal of Intelligent Systems and Applications* 9(4): 61.

Lee, W.-H., Jun, H.-G., and Kim, H.-J. 2015. "Hadoop Mapreduce performance enhancement using in-node combiners." arXiv preprint arXiv:1511.04861 *International Journal of Computer Science & Information Technology (IJCSIT)* 7(5): 1–17.

Narayan, S., Cohen, S. B., and Lapata, M. 2018. "Ranking sentences for extractive summarization with reinforcement learning." *arXiv e-prints*, arXiv-1802 (pp. 1747–1759).

Priya, V., and Umamaheswari, K. 2019a. "Aspect-based text summarization using MapReduce optimization." In Anandakumar, H., Arulmurugan, R., and Onn, C. (eds), *Computational Intelligence and Sustainable Systems. EAI/Springer Innovations in Communication and Computing* (pp. 131–139). Cham: Springer.

Priya, V., and Umamaheswari, K. 2019b. "Enhanced continuous and discrete multi objective particle swarm optimization for text summarization." *Cluster Computing* 22(1): 229–240.

Sadollah, A., Eskandar, H., Bahreininejad, A., and Kim, J. H. 2015. "Water cycle algorithm with evaporation rate for solving constrained and unconstrained optimization problems." *Applied Soft Computing* 30: 58–71.

Saini, N., Saha, S., Jangra, A., and Bhattacharyya, P. 2018. "Extractive single document summarization using multi-objective optimization: Exploring self-organized differential evolution, grey wolf optimizer and water cycle algorithm." *Knowledge-Based Systems* 164: 45–67.

Sanchez-Gomez, J. M., Vega-Rodríguez, M. A., and Pérez, C. J. 2018. "Extractive multi-document text summarization using a multi-objective artificial bee colony optimization approach." *Knowledge-Based Systems* 159: 1–8.

SAMPLE CODE

```
//ASPECTIDENTIFICATION//
importjava.io.File; importjava.io.FileOutputStream; importjava.io
  .PrintStream; importjava.util.ArrayList; importjava.util.Arrays;
  importjava.util.Collections; importjava.util.Enumeration;
  importjava.util.Hashtable; public class OpinionExtraction{
public static void main(String[] args) {
// processStopWords();
// processPosTag();
// extractOption(); AspectExtraction.classifyAspect();
File listData[] = new File("dataset").listFiles(); for (File current
  : listData) {
classify("aspects" + current.getName());
}
}
public static void processStopWords() { try {
File listData[] = new File("dataset").listFiles(); for (File current
  : listData) {
StopWordRemvoal.getStopword(current.getAbsolutePath());
}
} catch (Exception e) { e.printStackTrace();
}
}
//PSO SUMMARY// importjava.awt.Color;
importjava.awt.Dimension; importjava.awt.Toolkit; importjava.awt
  .event.ActionEvent; importjava.awt.event.ActionListener;
  importjava.io.File; importjava.io.FileInputStream; importjava.io
  .PrintStream; importjava.util.ArrayList; importjavax.swing
  .BorderFactory; importjavax.swing.ImageIcon; importjavax.swing
  .JButton; importjavax.swing.JInternalFrame; importjavax.swing.JPan
  el; importjavax.swing.JScrollPane; importjavax.swing.JTextArea;
  importjavax.swing.border.Border; importjavax.swing.border
  .CompoundBorder; importjavax.swing.border.EtchedBorder;
public class GASummeryForm extends JInternalFrame {
private static final long serialVersionUID = 6383459572274274626L;
  publicPSOSummeryForm() {
toInitialize(); setInitialize();
}
        publicPSOSummeryForm(String path) {
}
private void setScreenSize() {
Dimension localSysDimension = Toolkit.getDefaultToolkit().getScre
  enSize(); setBounds((localSysDimension.width - 900) / 2,
  (localSysDimension.height - 500) / 2, 900, 500);
}
private void setInitialize() { setTitle("Data Summery"); setScreenSize();
  classifyAspect(); setVisible(true);
}
public void classifyAspect() { loadFile("gasummery");
}
public void loadFile(String path) { try{
filePath
        FileInformationForm.txtInforamtion.append("\r\nFile Reading " +
          path, null); File listData[] = new File(path).listFiles();
        for (File current : listData) {
        txtPane.append("\r\nFile Name : " + current.getAbsolutePath());
        txtPane.append("\r\n~~~~~~~~~~~~~~~~~~~~~~~~~~~~~~~~~~~~~~~~~~~~~~
           ~~~~~~~~~~~
        ~~~~~~~~~~");
        txtPane.append("\r\n" + getFileContent(current.getAbsolutePath()));
           txtPane.append("\r\n~~~~~~~~~~~~~~~~~~~~~~~~~~~~~~~~~~~~~~~~~~~~~~
           ~~~~~~~~~~~
        ~~~~~~~~~~");
```

```
        FileInformationForm.txtInforamtion.append("Done ", null);
        }
} catch (Exception ez) { ez.printStackTrace();
}
} catch (Exception ez) { ez.printStackTrace();
}
        }
        private String getFileContent(String path) { try {
        FileInputStreaminputData = new FileInputStream(path); byte[]
            dataByte = new byte[inputData.available()]; inputData.re
            ad(dataByte);
        inputData.close();
        return new String(dataByte);
        } catch (Exception e) { e.printStackTrace();
        }
        return null;
        }
        private void toInitialize() { JPanelpanelView = new JPanel();
          panelView.setLayout(null);
        Border borderEtched = BorderFactory.createEtchedBorder(EtchedBorder
          .RAISED, Color.cyan, Color.blue);
        panelView.setBorder(new CompoundBorder(borderEtched, borderEtched));
          txtPane = new JTextArea();
        scrollPane = new JScrollPane(txtPane);
        btnClose = new JButton("Close", new ImageIcon("icon/close.png"));
          btnClose.addActionListener(new ActionListener() {
        public void actionPerformed(ActionEvent e) { setVisible(false);
        }});
        panelView.add(scrollPane); panelView.add(btnClose); scrollPane.
          setBounds(10, 10, 850,370);
                btnClose.setBounds(385, 400, 100, 30); getContentPane().
                  setLayout(null); getContentPane().add(panelView);
                  panelView.setBounds(10, 10, 870, 455);
                }public static String filePath = ""; JScrollPanescrollPane;
                  JTextAreatxtPane; JButtonbtnClose;
```

6 Performance Evaluation of Large-Scale Summarization Systems

6.1 EVALUATION OF SUMMARIES

Standard datasets for the evaluation of summaries are available extensively in the text summarization research. Few benchmark datasets extensively used for reporting accuracies of text summarization systems are:

i. Document Understanding Conference (DUC) datasets [1] such as DUC 2001, DUC 2002, DUC 2003, DUC 2004, DUC 2005, DUC 2006, and DUC 2007

Document Understanding Conferences (DUC) was sponsored by the Advanced Research and Development Activity (ARDA). The conference series is run by the National Institute of Standards and Technology (NIST) to further progress in summarization and enable researchers to participate in large-scale experiments. They provide datasets from DUC 2000 to DUC 2007 for the participants.

NIST produced 60 reference sets, 30 for training and 30 for testing. Each set contained documents, per-document summaries, and multi-document summaries, with sets defined by different types of criteria such as event sets, opinion sets, etc. The major tasks defined were:

- Fully automatic summarization of a single newswire/newspaper document (article):

Given such a document, a generic summary of the document with a length of approximately 100 words (whitespace-delimited tokens) was created.

Thirty sets of approximately ten documents each were provided as system input for this task.

- Fully automatic summarization of multiple newswire/newspaper documents (articles) on a single subject:

Given a set of such documents, four generic summaries of the entire set with lengths of approximately 400, 200, 100, and 50 words (whitespace-delimited tokens) were created.

DOI: 10.1201/9781003371199-6

Thirty document sets of approximately ten documents each were provided as system input for this task.

- Exploratory summarization:

Investigate alternative problems within summarization, novel approaches to their solution, and/or specialized evaluation strategies.

In the final conference conducted in 2006/2007, the main task was defined as follows:

- System task: Given a DUC topic and a set of 25 documents relevant to the topic, create from the documents a brief, well-organized, fluent summary which answers the need for information expressed in the topic.
- The summary can be no longer than 250 words (whitespace-delimited tokens). Summaries over the size limit will be truncated. No bonus will be given for creating a shorter summary. No specific formatting other than linear is allowed.

Also there was an update task in all the conferences which gives some additional criteria for summary generation.

ii. CNN and Daily Mail news [2] datasets used for training and testing purpose when used with supervised approaches

6.1.1 CNN Dataset

This dataset contains the documents and accompanying questions from the news articles of CNN. There are approximately 90k documents and 380k questions.

6.1.2 Daily Mail Dataset

This dataset contains the documents and accompanying questions from the news articles of Daily Mail. There are approximately 197k documents and 879k questions.

iii. Gigaword dataset [3]

It is popularly known as GIGAWORD dataset and contains nearly 10 million documents (over 4 billion words) of the original English Gigaword Fifth Edition. It consists of articles and their headlines. It consists of headline generation on a corpus of article pairs from Gigaword consisting of around 4 million articles. There are two features: document: article and summary: headline.

iv. Opinosis dataset [4]

This dataset contains sentences extracted from user reviews on a given topic. Example topics are 'performance of Toyota Camry,' 'sound quality of ipod nano,' etc. In total there are 51 such topics with each topic having approximately 100 sentences (on average). The reviews were obtained from various sources – Tripadvisor (hotels), Edmunds.com (cars), and Amazon.com (various electronics). There are approximately 100 sentences.

v. TAC dataset [5]

Text Analytics Conference 2011 Summarization Track is a more recent competition run by NIST as well that comes with data.

The goal of the Summarization Track is to foster research on systems that produce short, coherent summaries of text. The 2011 Summarization Track has three tasks:

Guided Summarization: The goal of guided summarization is to encourage a deeper linguistic (semantic) analysis of the source documents instead of relying only on document word frequencies to select important concepts. The guided summarization task is to write a 100-word summary of a set of ten newswire articles for a given topic, where the topic falls into a predefined category. Participants (and human summarizers) are given a list of aspects for each category, and a summary must include all aspects found for its category. Summaries will be evaluated for readability, content, and overall responsiveness. (The guided summarization task was run in the TAC 2010 Summarization Track.)

Automatically Evaluating Summaries of Peers (AESOP): The AESOP task is to automatically score a summary for a given metric. It complements the basic summarization task by building a collection of automatic evaluation tools that support the development of summarization systems. (AESOP was run in the TAC 2010 Summarization Track.)

MultiLing Pilot: The aim of the MultiLing Pilot is to foster and promote the use of multilingual algorithms for summarization. This includes the effort of transforming an algorithm or a set of resources from a monolingual to a multilingual version.

Evaluation of summarization systems is a crucial part in designing automatic text summarization systems. This involves benchmarking the quality of the summary generated using automatic text summarization system against ground truth summary.

vi. News articles and review datasets

A collection of datasets are available for summarization task at Kaggle [6] repository. Some major collections are described.

vii. BBC news datasets [7]

6.1.3 Description

This dataset for extractive text summarization has 417 political news articles of BBC from 2004 to 2005 in the News Articles folder. For each article, five summaries are provided in the Summaries folder. The first clause of the text of articles is the respective title.

This dataset was created using a dataset used for data categorization that consists of 2225 documents from the BBC news website corresponding to stories in five topical areas from 2004 to 2005.

6.2 METHODOLOGIES

Evaluation measures for automatic text summarization systems [9] could be broadly categorized into intrinsic and extrinsic evaluation.

6.2.1 Intrinsic Methods

Intrinsic evaluation determines the summary quality on the basis of comparison between the automatically generated summary and the human-generated summary.

6.2.2 Extrinsic Methods

Extrinsic evaluation methods focus on the impact of summarization on other tasks such as document categorization, information retrieval, and question answering systems. Both the methods are explained in detail in the following sections.

6.3 INTRINSIC METHODS

Intrinsic methods directly assess the system-generated summary using standard measures. Reference summaries or gold standard summaries for comparison with system-generated summaries are created using human annotators. Manual summaries are created by human experts by understanding the text documents used as input for summarization. Commonly used measures for measuring the summary quality and content evaluation are listed in Table 6.1. The measures are explained as follows.

6.3.1 Text Quality Measures

6.3.1.1 Grammaticality

This measure gives information about grammatical features in the summary. It checks whether the text should not contain non-textual items (i.e. markers) or punctuation errors or incorrect words.

6.3.1.2 Non-redundancy

Non-redundancy refers to novelty in a summary, where in novelty denotes the minimum number of redundant sentences in the summary as shown in Equation 6.1.

$$\text{Novelty}\left(\text{Sum}_s\right)=1-\max\left(\text{sim}\left(S_x,S_y\right)\right) \tag{6.1}$$

TABLE 6.1
Text Summarization Measures

S.No	Measures	
1	Text quality measures	Grammaticality
2		Non-redundancy
3		Reverential clarity
4		Structure and coherence
5	Content evaluation	Precision, recall, and *F*-score
6		Relative utility
7		Cosine similarity
8		Unit overlap
9		Longest common subsequence
10		n-gram matching (ROUGE)
11		Pyramids
12		LSA-based measure

6.3.1.3 Reverential Clarity

This metric measures the references in the document. The references indicate unsolved references that remain as personal pronouns. This made the reader unable to find out what is involved in the sentence. So, to generate a good summary, we have to replace the first mention of a reference in the summary. Some researchers found that the introduction of anaphoric resolution can slightly improve the readability of the summary. Also, the authors confirm that the impact of anaphora resolution (AR) varies from one domain to another which means it improves some domain summaries compared to others.

6.3.1.4 Structure and Coherence

Coherence is represented in terms of relations between text segments. This is generally represented based on the structure of the text. This could be analyzed using lexical chains and anaphora resolution.

6.3.2 Co-selection-Based Methods

Co-selection measures can count as a match with exactly the same sentences between system summaries and reference summaries.

6.3.2.1 Precision, Recall, and *F*-score

The main evaluation metrics of co-selection are precision, recall, and F-score. Precision (*P*) is the number of sentences occurring in both system and ideal summaries divided by the number of sentences in the system summary. Recall (*R*) is the number of sentences occurring in both system and ideal summaries divided by the number of sentences in the ideal summary. *F*-score is a composite measure that combines precision and recall. The basic way to compute the F-score is to count a harmonic average of precision and recall and is shown in Equation 6.2.

$$F = \frac{2PR}{P+R} \tag{6.2}$$

Formula for measuring the F-score is shown in Equation 6.3.

$$F = \frac{\left(\beta^2+1\right)PR}{\beta^2 P+R} \tag{6.3}$$

where β is a weighting factor that favors precision when $\beta > 1$ and favors recall when $\beta < 1$.

6.3.2.2 Relative Utility

The relative utility (RU) measure was introduced by Radev et al. (2000) [12]. With RU, the model summary represents all sentences of the input document with confidence values for their inclusion in the summary. For example, a document with five sentences [1 2 3 4 5] is represented as [1/5 2/4 3/4 4/1 5/2]. The second number in each pair indicates the degree to which the given sentence should be part of the summary according to a human judge. This number is called the utility of the sentence. It depends on the input document, the summary length, and the judge. In the example, the system that selects sentences [1 2] will not get a higher score than a system that chooses sentences [1 3] because both the summaries [1 2] and [1 3] carry the same number of utility points (5 + 4). Given that no other combination of two sentences carries a higher utility, both the systems [1 2] and [1 3] produce optimal extracts. To compute relative utility, a number of judges ($N \geq 1$) are asked to assign utility scores to all N sentences in a document. The top e sentences according to utility score 3 are then called a sentence extract of size e. Then the following system performance metric RU can be given as per Equation 6.4:

$$\text{RU} = \frac{\sum_{j=1}^{n} \delta_j \sum_{i=1}^{N} u_{ij}}{\sum_{j=1}^{n} \varepsilon_j \sum_{i=1}^{N} u_{ij}} \tag{6.4}$$

where u_{ij} is a utility score of sentence j from annotator i, ε_j is 1 for the top e sentences according to the sum of utility scores from all judges, otherwise its value is 0, and δ_j is equal to 1 for the top e sentences extracted by the system, otherwise its value is 0.

6.3.3 Content-Based Methods

6.3.3.1 Content-Based Measures

Content-based measures capture the fact that two sentences can contain the same information even if they are written differently. Furthermore, summaries written by two different annotators do not in general share identical sentences.

6.3.3.2 Cosine Similarity

Similarity between system summary and its reference summary is measured using cosine similarity. This metric is given by the formula Equation 6.5.

$$\cos(X,Y) = \frac{\sum_i x_i . y_i}{\sqrt{\sum_i (x_i)^2} . \sqrt{\sum_i (y_i)^2}} \tag{6.5}$$

where X and Y represent system-generated summary and reference summary. They are represented using vector space model.

6.3.3.3 Unit Overlap

Unit overlap is a metric that measures the overlap of word units between system summary and reference summary. The formula is given in Equation 6.6

$$\text{overlap}(X,Y) = \frac{X \cap Y}{X + Y - X \cap Y} \tag{6.6}$$

where X and Y are representations based on sets of words or lemmas. $\|X\|$ is the size of set X.

6.3.3.4 Longest Common Subsequence

This metric is given by Radev et al. (2015). This measures the sequence of words and the length of the longest common subsequence as in Equation 6.7.

$$\text{lcs}(X,Y) = \frac{\text{length}(X) + \text{length}(Y) - \text{edit}_{di}(X,Y)}{2} \tag{6.7}$$

where X and Y are representations based on sequences of words or lemmas, lcs(X, Y) is the length of the longest common subsequence between X and Y, length(X) is the length of the string X, and $\text{edit}_{di}(X, Y)$ is the edit distance of X and Y.

6.3.3.5 N-Gram Co-occurrence Statistics: ROUGE

Recall-Oriented Understudy for Gisting Evaluation (ROUGE) was used as an automatic evaluation method. The ROUGE family of measures, which are based on the similarity of n-grams, was introduced in 2004 by Lin et al. Suppose a number of annotators created reference summaries – reference summary set (RSS). The ROUGE-n score of a candidate summary is computed using Equation 6.8:

$$\text{ROUGE} - n = \frac{\sum_{c \in \text{RSS}} \sum_{\text{gram}_{n \in c}} \text{Count}_{\text{match}}(\text{gram}_n)}{\sum_{\text{gram}_{n \in c}} \text{Count}(\text{gram}_n)} \tag{6.8}$$

where $\text{Count}_{\text{match}}(\text{gram}_n)$ is the maximum number of n-grams co-occurring in a candidate summary and a reference summary and $\text{Count}(\text{gram}_n)$ is the number of n-grams in the reference summary. The average n-gram ROUGE score, ROUGE-n, is a recall-based metric as already described in Section 6.3.2.

6.3.3.6 Pyramids

The pyramid method is a novel semiautomatic evaluation method by Nenkova and Passonneau (2005). Its basic idea is to identify summarization content units (SCUs) that are used for the comparison of information in summaries. SCUs emerge from the annotation of a corpus of summaries and are not bigger than a clause. The annotation starts with identifying similar sentences and then proceeds with fine-grained inspection that can lead to identifying related subparts more tightly. SCUs that appear in more manual summaries will get greater weights, so a pyramid will be formed after SCU annotation of manual summaries. At the top of the pyramid, there are SCUs that appear in most of the summaries, and thus they have the greatest weight. The lower in the pyramid the SCU appears, the lower its weight is because it is contained in fewer summaries. The SCUs in peer summary are then compared against an existing pyramid to evaluate how much information agrees between the peer summary and manual summary.

6.3.3.7 LSA-Based Measure

The ability to capture the most important topics is used by the two evaluation metrics by Yeh et al. (2005). The idea is that a summary should contain the most important topic(s) of the reference document (e.g. full text or abstract). It evaluates a summary quality via content similarity between a reference document and the summary like other content-based evaluation measures do. Singular value decomposition (SVD) performs a breakdown which represents the degree of term importance in salient topics. The methods measure the similarity between the matrix U derived from the SVD performed on the reference document and the matrix U derived from the SVD performed on the summary. To appraise this similarity, two measures have been proposed: (i) main topic similarity and (ii) term significance similarity.

6.3.3.8 Main Topic Similarity

The main topic similarity measure compares the first left singular vectors of the SVD performed on the reference document, which are main topics, and the SVD performed on the summary. These vectors correspond to the most important word pattern in the reference text and the summary. This is called as the main topic. The cosine of the angle between the first left singular vectors is measured.

6.3.3.9 Term Significance Similarity

The LSA measure compares a summary with the reference document from an angle of r vector with most salient topics. The idea behind it is that there should be the same important topics/terms in both the documents. The first step is to perform the SVD on both the reference document and the summary matrices. Then to reduce the dimensionality of the documents, SVDs are performed to extract only the important topics there.

6.4 EXTRINSIC METHODS

6.4.1 Document Categorization

The quality of automatic summaries can be measured by their suitability for surrogating full documents for categorization. Here the evaluation seeks to determine whether the generic summary is effective in capturing whatever information in the document is needed to correctly categorize the document. A corpus of documents together with the topics they belong to is needed for this task. Results obtained by categorizing summaries are usually compared to those obtained by categorizing full documents (upper bound) or random sentence extracts (lower bound). Categorization can be performed either manually by Mani et al. (1999) or by a machine classifier by Hainek et al. (2003). If we use an automatic categorization, we must keep in mind that the classifier demonstrates some inherent errors. It is therefore necessary to differentiate between the error generated by a classifier and the one by a summarizer. It is often done only by comparing the system performance with the upper and lower bounds. The common strategy is to use relevancy as detailed below.

Given a document, which could be a generic summary or a full-text source (the subject was not told which), the human subject chose a single category (from five categories, each of which had an associated topic description) to which the document is relevant, or else chose 'none of the above.' Precision and recall of categorization are the main evaluation metrics. Precision in this context is the number of correct topics assigned to a document divided by the total number of topics assigned to the document. Recall is the number of correct topics assigned to a document divided by the total number of topics that should be assigned to the document. The measures go against each other, and therefore, a composite measure F-score can be used.

6.4.1.1 Information Retrieval

Information retrieval (IR) is another task appropriate for the task-based evaluation of a summary quality. Relevance correlation given by Radev et al. (2003) is an IR-based measure for assessing the relative decrease in retrieval performance when moving from full documents to summaries. If a summary captures the main points of a document, then an IR machine indexed on a set of such summaries (instead of a set of the full documents) should produce (almost) a better result. Moreover, the difference between how well the summaries do and how well the full documents do should serve as a possible measure for the quality of summaries.

Suppose that given query Q and a corpus of documents D, a search engine ranks all the documents in D according to their relevance to query Q. If instead of corpus D, the corresponding summaries of all the documents are substituted for the full documents and the resulting corpus of summaries S is ranked by the same retrieval engine for relevance to the query, a different ranking will be obtained. If the summaries are good surrogates for the full documents, then it can be expected that the ranking will be similar. There exist several methods for measuring the similarity of rankings. One such method is Kendall's tau and another is Spearman's rank correlation given by Seigel et al. (1998). However, since search engines produce relevance scores in addition to rankings, we can use a stronger similarity test, linear correlation. Relevance correlation (RC) is defined as the linear correlation of the relevance scores assigned by the same IR algorithm in different datasets.

6.4.1.2 Question Answering

An extrinsic evaluation of the impact of summarization in a task of question answering was carried out by Morris et al. (1992). The authors picked four Graduate Management Admission Test (GMAT) reading comprehension exercises. The exercises were multiple choices, with a single answer to be selected from answers shown alongside each question. The authors measured how many of the questions the subjects answered correctly under different conditions. They were shown the original passages, then an automatically generated summary, furthermore a human abstract created by a professional abstractor instructed to create informative abstracts. Finally, the subjects had to pick the correct answer just from seeing the questions without seeing anything else. The results of answering in different conditions were then compared.

There are other methods like similarity detection for summary evaluation proposed by Priya et al. (2019). This method identifies the similarity between textual summaries generated by two summarization systems. This similarity detection uses graphical links between the documents based on their semantic knowledge bases.

Implications:

Large-scale evaluations are simple measures like precision, recall, and percent agreement which (i) do not take casual agreement into account and (ii) do not account for the fact that human judges don't approve which sentences should be in a summary. They are expensive (an approach using manual judgments can scale up to a few hundred summaries but not to tens or hundreds of thousands).

6.4.2 Summary

This chapter gives a detailed description of two major evaluation measures: intrinsic and extrinsic evaluation. All the measures are given with detailed formula and other methods used for evaluation. It also discusses about the challenges in large-scale summarization. This also gives elaborate information about the benchmarked datasets used for text summarization.

6.4.3 Examples

1. Given the following bigrams in each summary, compute ROUGE scores for precision, recall, and *F*-score.
   Manual Summary: ArgR has N-terminal and C-terminal domains
   Automatic Summary: ArgR has N-terminal domain

MS: ('**ArgR has**', '**has N-terminal**', 'N-terminal and', 'and C-terminal', 'C-terminal domains'); $m=5$
AS: ('**ArgR has**', '**has N-terminal**', 'N-terminal domain'); $n=3$

$$\text{Recall} = \frac{2-\text{Gram}(\text{MS,AS})}{|m|}$$

$$\text{Precision} = \frac{2-\text{Gram}(\text{MS,AS})}{|n|}$$

$$\text{F}-\text{Score} = \frac{2^*\text{Recall}^*\text{Precision}}{\text{Recall}+\text{Precision}}$$

There are two overlaps: 'ArgR has' and 'has N-terminal.' The ROUGE-2 scores for AS are:

Recall = 2/5 = 0.4, precision = 2/3 = 0.6, and *F*-score = 0.5.

2. Use the same summaries as given in exercise 1 and compute ROUGE-SU4.

ROUGE-SU4 measures the co-occurrence of unigrams together with skip bigrams of distance 4, that is, all pairs of words separated at most by four words (following the sentence order). Let SKIP4 (MS, AS) be the overlapped unigrams and skip bigrams; then we can calculate ROUGE-SU4 scores as shown in the equation below:

$$\text{Recall} = \frac{\text{skip4}(\text{MS,AS})}{|m|}$$

$$\text{Precision} = \frac{\text{skip4}(\text{MS,AS})}{|n|}$$

MS: ('**ArgR**', 'has', '**N-terminal**', 'and', 'C-terminal', 'domains') + ('ArgR has', 'ArgR N-terminal', 'ArgR and', 'ArgR C-terminal', 'has N-terminal', 'has and', 'has C-terminal', 'has domains', 'N-terminal and', 'N-terminal C-terminal', 'N-terminal domains', 'and C-terminal', 'and domains', 'C-terminal domains'); $m=20$.
AS': ('domain', '**N-terminal**', 'of', '**ArgR**',) + ('domain N-terminal', 'domain of', 'domain ArgR', 'N-terminal of', 'N-terminal ArgR', 'of ArgR'); $n=10$.

Then, based on the following unigrams and skip bigrams of the manual summary, MS, and the automatic summary, AS', the obtained ROUGE-SU4 scores are:

Recall = 2/20 = 0.1, precision = 2/10 = 0.2, and *F*-score = 0.1, because there are only two coinciding unigrams. This score is low, but it takes into consideration isolated words, and the effect of word order is reduced.

3. Consider a dataset with a 1:100 minority-to-majority ratio, with 100 minority examples and 10,000 majority class examples.

   A model makes predictions and predicts 120 examples as belonging to the minority class, 90 of which are correct, and 30 of which are incorrect.

   The precision for this model is calculated as:

   Precision = True Positives / (True Positives + False Positives)

   Precision = 90 / (90 + 30)

   Precision = 90 / 120

   Precision = 0.75

The result is a precision of 0.75, which is a reasonable value but not outstanding. You can see that precision is simply the ratio of correct positive predictions out of all positive predictions made, or the accuracy of minority class predictions.

4. Consider the same dataset, where a model predicts 50 examples belonging to the minority class, 45 of which are true positives and five of which are false positives. We can calculate the precision for this model as follows:

   Precision = True Positives / (True Positives + False Positives)

   Precision = 45 / (45 + 5)

   Precision = 45 / 50

   Precision = 0.90

Exercises:

1. A test data consisted of 91 data points. We also notice that there are some actual and predicted values. The actual values are the number of data points that were originally categorized into 0 or 1. The predicted values are the number of data points predicted by KNN model as 0 or 1.

   The actual values are:

   The patients who actually don't have a heart disease = 41

   The patients who actually do have a heart disease = 50

The predicted values are:

   Number of patients who were predicted as not having a heart disease = 40

   Number of patients who were predicted as having a heart disease = 51

   Calculate precision, recall, and *F*-measure using the data given.

2. Explore this notion by looking at the following figure, which shows 30 predictions made by an email classification model. Those to the right of the classification threshold are classified as 'spam,' while those to the left are classified as 'not spam.'

   Now calculate precision and recall based on the results shown in Figure 6.1.

3. Consider the summaries given below:

   System-generated summary 1:

   the cat was found under the bed

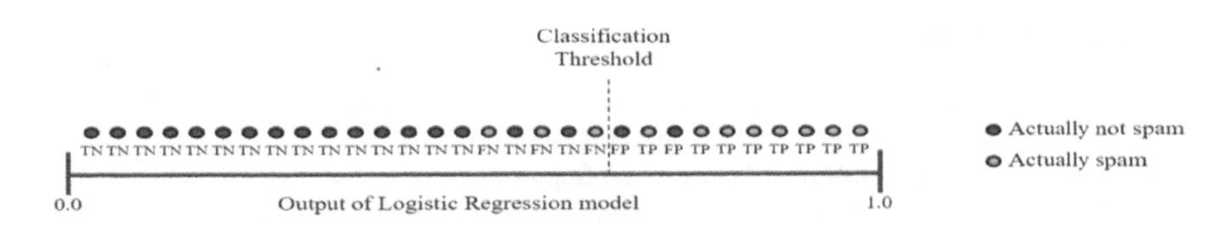

FIGURE 6.1 Exercise 2

System-generated summary 2:
the tiny little cat was found under the big funny bed
Manual/gold standard summaries:
the cat was under the bed
Calculate precision, recall, and *F*-measure for ROUGE metrics for system summary 1 and 2.

Note: Use the formulas given:

$$\text{Recall} = \frac{\text{Number of overlapping words}}{\text{Total words in reference summary}}$$

$$\text{Precision} = \frac{\text{Number of overlapping words}}{\text{Total words in system summary}}$$

$$\text{F} - \text{Score} = \frac{2 * \text{Recall} * \text{Precision}}{\text{Recall} + \text{Precision}}$$

4. Compute ROUGE-2 metrics based on the hint below for the given text summaries.

   Hint: ROUGE-1 refers to the overlap of **unigrams** between the system and reference summaries. **ROUGE-2** refers to the overlap of **bigrams** between the system and reference summaries.
   System-generated summary 1:
   the cat was found under the bed
   Manual/gold standard summaries:
   the cat was under the bed
5. Compute ROUGE metrics for the following data:
   (i)
   Reference summary – The dog bites the man.
   System summary – The man was bitten by the dog, and find in dark.
   (ii)
   System summary = [["Tokyo is the one of the biggest city in the world."]]
   Reference summary = [[["The capital of Japan, Tokyo, is one of the b iggest cities of the world."]]]

BIBLIOGRAPHY

https://www.kaggle.com/datasets/disisbig/hindi-text-short-summarization-corpus

http://duc.nist.gov/data.html (Accessed on September 20, 2020)

http://cs.nyu.edu/~kcho/DMQA/ (Accessed on September 20, 2020)

https://catalog.ldc.upenn.edu/ldc2003t05 (Accessed on September 20, 2020)

https://github.com/kavgan/opinosis-summarization/blob/master/OpinosisDataset1.0_0 .zip

https://tac.nist.gov//2011/Summarization/

https://www.kaggle.com/

http://mlg.ucd.ie/datasets/bbc.html

Hynek, J., and Jeˇzek, K. 2003. "Practical approach to automatic text summarization." In *Proceedings of the ELPUB'03 Conference*, Guimaraes, Portugal (pp. 378–388).

Joshi, A., Fernández, E. F., and Alegre, E. 2018. "Deep learning based text summarization: Approaches databases and evaluation measures." In *International Conference of Applications of Intelligent Systems.*, ios,pp1–4

Lin, C. 2004. "ROUGE: A package for automatic evaluation of summaries." In *Proceedings of the Workshop on Text Summarization Branches Out*, Barcelona, Spain: ACL (pp 74–81).

Mani, I., Firmin, T., House, D., Klein, G., Sundheim, B., and Hirschman, L. 1999. "The TIPSTER summac text summarization evaluation." In *Proceedings of the 9th Meeting of the European Chapter of the Association for Computational Linguistics* (pp. 77–85).

Morris, A., Kasper, G., and Adams, D. 1992. "The effects and limitations of automatic text condensing on reading comprehension performance." *Information Systems Research* 3(1): 17–35.

Nenkova, A., and Passonneau, R. 2004. "Evaluating content selection in summarization: The pyramid method NAACL-HLT." pp 1–8.

Priya, V., and Umamaheswari, K. 2019. "A document similarity approach using grammatical linkages with graph databases." *International Journal of Enterprise Network Management* 10(3–4): 211–223.

Radev, D. R., Jing, H., Styś, M., and Tam, D. 2004. "Centroid-based summarization of multiple documents." *Information Processing & Management* 40(6): 919–938.

Radev, D., Teufel, S., Saggion, H., Lam, W., Blitzer, J., Qi, H., Celebi, A., Liu, D., and Drabek, E. 2003, July. "Evaluation challenges in large-scale document summarization." In *Proceedings of the 41st Meeting of the Association for Computational Linguistics*, Sapporo, Japan(pp. 375–382).

Siegel, S., and Castellan, N. J. 1988. *Nonparametric Statistics for the Behavioral Sciences*, 2nd edn. Berkeley, CA: McGraw-Hill.

Spark-Jones, K., and Galliers, J. R. 1995. "Evaluating natural language processing systems: An analysis and review." Number 1083 in *Lecture Notes in Artificial Intelligence.* Heidelberg: Springer.

Steinberger, J., and Ježek, K. 2012. "Evaluation measures for text summarization." *Computing and Informatics* 28(2): 251–275.

Yeh, J. Y., Ke, H. R., Yang, W. P., and Meng, I. H. 2005. "Text summarization using a trainable summarizer and latent semantic analysis." *Information Processing and Management* 41(1): 75–95.

7 Applications and Future Directions

7.1 POSSIBLE DIRECTIONS IN MODELING TEXT SUMMARIZATION

Text summarization has been modeled using several techniques. Significant among them are:

i. Deep learning models
ii. Topic modeling methods
iii. Evolutionary models
iv. Graph models
v. Classification model
vi. Clustering model

Some main challenges in text modeling for summarization are:

- Different size of corpora and vocabulary makes modeling text summarization difficult.
- Complex structure of language and semantic representations are challenging for new languages.
- Long-term structure of language or sequencing based on language dictionaries becomes tough in text modeling.

7.2 SCOPE OF SUMMARIZATION SYSTEMS IN DIFFERENT APPLICATIONS

Summarization systems play a vital role in several domains. They are:

i. Legal cases (rent control, income tax, and sales tax)
ii. News articles
iii. Bio-medical domain
iv. Scientific papers
v. Social media posts
vi. Web content–based system
vii. E-mail contents
viii. Healthcare domain
ix. Cognitive domain (audio/video)
x. Communication industry

DOI: 10.1201/9781003371199-7

Few domains which find extensive usage for summarization have been discussed in detail in the following sections.

7.3 HEALTHCARE DOMAIN

Medical information is distributed through various levels such as:

- Public health information systems
- Clinical information systems
- Imaging systems using radiology
- Systems involving molecular and cellular processes

These systems consist of various document types: scientific articles, electronic medical records, semi-structured databases, web documents, e-mailed reports, X-ray images, and videos. The characteristics of all the document types have to be taken into account in developing summarization systems.

Scientific articles are mainly composed of text, and in specific they have multiple sections that can be explored by a summarization system. The number of scientific journals in the fields of health and biomedicine is extremely large for even a single specialty. This makes it difficult for physicians and researchers to explore the latest results reported in their fields. Scientific articles may contain structured data like tables, graphs, or images apart from the usual text.

Electronic medical records contain structured data, apart from free text. There are several online databases, which contain the abstracts and citation information of most articles in the medicine field. One such online database is MEDLINE, which contains abstracts from more than 3500 journals. MEDLINE provides keyword search and returns the corresponding abstracts that contain those keywords. In the database, abstracts are indexed according to the Medical Subject Headings (MeSH) thesaurus. Apart from access to the abstracts, MEDLINE also provides full citations of the articles along with links to the articles themselves, in case they are online. Web documents may appear in health directories and catalogs, which need to be searched first in order to locate interesting web pages. The web page layout is also another factor that needs to be taken into account.

E-mailed reports are mainly free text without any other structure. ProMED-mail is a free public service, which encourages the exchange of information regarding epidemics. There are non-free services like MDLinx, which provide physicians and researchers to subscribe and receive alerts regarding new findings in their specialty fields. The use of e-mailed reports for fast distribution of epidemiological information by the Internet shows increasing success in monitoring epidemiological events.

X-ray images and videos using echocardiograms represent a completely different document type. They are of multimedia in nature. Such documents can be graphs, such as cardiograms; images, such as X-rays, etc.; and videos, such as the various echograms, e.g. echocardiograms, echoencephalograms. Medical videos are used mainly for educational purposes, e.g. videos of clinical operations or videos of dialogs between the doctor and the patient. Most of these documents are now transcribed and stored in digital form. This is a completely different type of medical documents, which may contain very remarkable information that should be added to a summary. Compared to other domains, medical documents have certain unique characteristics that make their analysis very challenging and attractive. Some of the unique characteristics of medical documents are that they contain both comprehensive and accurate information. The uniqueness of medical documents is due to their volume and their heterogeneity.

Significant summarization techniques used in the medical domain are:

i. Abstractive single-document summarization
ii. Extractive multi-document summarization
iii. Multimedia summarization
iv. Cognitive model–based summarization

Future Directions for Medical Document Summarization

Some common challenges are scaling of large-size collection of documents and annotated summarization corpora for training and testing. There are wide ranges of challenges in summarization tasks in the medical domain. They are:

Input type: Medical domain can provide a lot of useful inputs in several media such as speech, images, and videos. Summarizing relevant data from different information such as spoken transcriptions and textual reports related to the specific echo videos is an important issue for practical applications. This is an encouraging direction for future research and development.

Language of the input documents: There are a few multilingual aspects that are taken into account in the medical domain other than English. There are many resources available in several other languages and also tools that can be explored in building summarizers for handling more than one language.

User-oriented summaries: There are different user types like clinicians, researchers, patients, and specific users. The users' preferences should be taken into account along with their expertise in the domain, as well as the users' access tasks. The summary appearance can also be affected by the user's model in the form of text and images. Personalized access to medical information is a crucial issue and needs to be further investigated.

Domain customization is another significant issue. Importance must be given to the growth of technology that can be easily ported to new subdomains. A general-purpose system must be able to explore the various characteristics of medical documents. Sectioning of scientific articles and the specialized language used in e-mailed reports or in-patient records are important features. These features can significantly affect the performance of the involved language processing tools.

Output factor: The quality of the summarization output is strongly related to the summarization task. Criteria for qualitative and quantitative summarization need to be established based on domain knowledge and the users' interests.

7.4 SOCIAL MEDIA

Social media posts and blogs found numerous applications in the text summarization field. There are some major types of social media platforms. A description of the types and their data are given below:

- Social networks:

Social networks are also called relationship networks. These networks help people and organizations connect online to share information and ideas. Examples: Facebook, Twitter, LinkedIn.

The data in the networks include text, photos, comments, videos, and audio files. These unstructured data can be used for summarization. All the data in social networks pertains to some information which holds some value. Gupta et al. (2019) briefed a multilingual text summarization system using Twitter data. This uses a machine translation technique, and specific content-based summarization was performed. Topic summarization by Duan et al. (2012) has also been effective using subtopics. This generates a short and compact summary for a collection of tweets on the same or similar topics. Event detection, event summarization, and safe route recommendation are a few applications of Twitter data summarization.

- Media sharing networks

In general people use these networks to find and share photos, videos, live videos, and other media online. Examples: Instagram, Snapchat, YouTube. Video summarization is a separate domain as the content and the methodology are different compared to text summarization. Aswin et al. (2019) proposed subtitle generation based on video summarization. They used natural language processing (NLP) and speech recognition techniques.

- Discussion forums

People use this community question answering and discussion forums to share, discuss, and share news, information, and opinions. These are one of the oldest types of social media. Examples: Reddit, Quora, Digg, StackOverflow. In discussion forum threads, summarization (2018) can be useful for the selection of the best answer, sharing opinions and long forum threads on a single topic for finding the impact of any event. In this work, summarization had been done by user evaluation study and agreement between human raters.

- Bookmarking and content curation networks

People use this to discuss new and trending content in media. Bookmarking and content-based networks help people discover, save, and share the content and media. Examples: Pinterest, Flipboard.

Cong et al. (2017) solved web media summarization problems: Building a new dataset consisting of around 500 users, 3000 albums, and 1 million images and modeled the dictionary selection issue as a video summarization problem. This can be used for automatic album creation and titling.

- Consumer review networks

People use this to find, review, and share information about brands, products, and services, as well as restaurants, travel destinations, and more. Consumer review networks give people a place to review brands, businesses, products, services, travel spots, and just about anything else. Examples: Yelp, Zomato, Tripadvisor Review summarization by many researchers helps to find trends on recent topics and users' opinions on various services.

- Blogging and publishing networks

People use these networks to publish, discover, and comment on content online. Examples: WordPress, Tumblr, Medium.

Maylawati et al. (2019) combine sequential pattern mining (SPM) and deep learning (DL) for a better text summarization process. Sequence of words is a text representation used in this research with sequential pattern mining.

- Social shopping networks

People use shopping networks to spot trends, follow brands, share great finds, and make purchases. Examples: Polyvore, Etsy, Fancy. This e-commerce given by Turban et al. (2018) can provide valuable services to all stakeholders using summarization. Some of them are trend prediction, sale prediction, product recommendation, etc.

- Interest-based networks

People use these interest-based networks to connect with others around a shared interest or hobby in the reading community. Interest-based networks take a more targeted approach than the big social networks do by focusing solely on a single subject, such as books, music, or home design. Examples: Goodreads, Houzz, Last.fm.

This network review summarization given by Wang et al. (2019) is useful in applications like book impact assessment, information behaviors, users' behaviors, and recommendation.

- Sharing economy networks

People use these economy networks to advertise, find, share, buy, sell, and trade products and services between peers. They are also called 'collaborative economy networks,' which connect people online for various purposes. Some of them are advertising, finding, sharing, buying, selling, and trading products and services. Examples: Airbnb, Uber, TaskRabbit.

Sharing economy firms such as Uber and Airbnb facilitate trusted transactions between strangers on digital platforms. This creates economic and other values but raises concerns around racial bias, safety, and fairness to competitors. The work given by Cheeks et al. (2017) utilizes advanced machine learning for an automatic understanding of implicit structures. This work uses a new concept of double subjectivity for large-scale machine learning of unstructured text.

- Anonymous social networks

People use these anonymous networks to gossip, vent, snoop, and sometimes bully. Examples: Whisper, Ask.fm, After School. Pudaruth et al. (2016) framed an intelligent question answering system using summarization for ask.com reviews or answers. Summarization of these data would be helpful in the identification of fake sites, fake reviews or gossips, and so on.

Challenges in Social Media Text Summarization

Domain Knowledge and Transfer Learning

Robust machine learning techniques tested in different types of disasters such as floods and hurricanes have yet to be developed for summarization. Methods using domain knowledge and transfer learning can be employed to utilize event data summarization from past events, possibly in combination with unlabeled from the new event.

Online Learning

The presence of false messages in the training and testing sets may lead to misleading summaries. New redundancy removal strategy techniques to uncover new concepts and to track concept evolution in the data stream can be used. This can help in detecting underlying data changes and trigger model retraining processes.

Information Credibility

Fake review detection, dealing with misinformation cases, and determining information credibility are new challenges regarding information credibility. Finding the credibility of information on social media is a challenging task. Methods for the identification of any bias associated with the social data and considering a variety of different sources are key challenges regarding information credibility.

Applications of Deep Learning

Although deep neural network techniques are promising for social media data summarization, they require large amounts of training data. Recent attempts to make crisis-related labeled data available are hopeful for summarization, but much more remains to be done.

IMPLICIT AND EXPLICIT INFORMATION FOR ACTIONABLE INSIGHTS

Despite extensive research that mainly focuses on understanding and extracting updates contributing to both implicit and explicit information from social media platforms is also significantly increasing. In this perspective, limited work has been done that focuses on understanding this information in a given context which is useful for the stakeholders. This could be user-oriented summarization. Summarization systems that categorize summaries to match different information needs of organizations become essential for solving novel requirements.

7.5 RESEARCH DIRECTIONS FOR TEXT SUMMARIZATION

Extractive summarization systems aim at selecting the most relevant sentences from the collection of original documents in order to produce a condensed text rendering important pieces of information. There are a few open limitations which remain unsolved in automatic summarization systems. They are:

Particularly in multi-document summarization,

- The selection of sentences from different documents leads to redundancy, which in turn must be eliminated.
- In summary document only part of a sentence or partial sentence is relevant. So extraction of sub-sentences could lead to maximum relevancy.
- Extraction of sentences from different documents may produce an inconsistent and/or not readable summary.

These limitations lead to some active research directions. They are:

- Redundancy:

In order to overcome the redundancy problem, researchers are actively working on a better representation of the text content. Summaries tailored towards specific user needs are being researched. The combination of new techniques and innovative algorithmic modules has been proven to be necessary for a more fine-grained representation of the original set of documents (e.g. the integration of an opinion-mining module for opinion-based summarization).

- Sentence compression:

This is another very active domain of research. Sentence compression aims at keeping only the part of the sentence that is really meaningful and of interest for the abstract. Most often, compression rules are defined manually. Some recent experiments tried to automatically learn these rules from a set of examples. The improvement brought by this method in readability and summarization quality still needs to be assessed. Most approaches using sentence compression require an accurate analysis of the input sentence in order to provide reliable results. So, sentence compression is a promising source of improvement, but its application still needs to be validated.

- Readability:

The main drawback of extractive methods is the lack of readability of the text produced. In most systems, sentence ordering is based on simple heuristics that are not adequate to create a coherent text. Recent research aims at finding new methods for producing more coherent texts. For example recommend that computing the local coherence among candidate sentences may lead to more readable summaries. Developing a global document model may also help in improving readability.

- Evaluation of large-scale summaries:

The evaluation of automatic summarization is an open issue. Recent automatic methods like Pyramid have been proven more consistent than human-based evaluation. The real problem is the lack of agreement between humans when evaluating summaries. The development of more focused summaries may lead to a more reliable evaluation. This also leads to a better convergence between human and automatic evaluation methods. As a whole, large-scale automatic summarization evaluation is still a promising research area with several challenges ahead.

Other major issues of summarization are:

i. Extraction of the important sentences and sentence ordering.
ii. Sentences in the summary have to be in the order as in the source document for producing an efficient summary.
iii. Compressions involving lexical substitutions, paraphrasing, and reformulation components could be added to improve the quality of the summary.
iv. Addition of these components would make the automatic summarization systems to become more complex.
v. The capability of the system is inhibited by the richness of their representation, and their way to create such structure is the greatest challenge in text summarization.
vi. Still information diffusion is not handled properly using text summarization.
vii. Word sense ambiguity, this is the ambiguity created sometimes due to abbreviations that can have more than one acronym.
viii. Interpretation of long sentences and jargons.

7.6 FURTHER SCOPE OF RESEARCH ON LARGE-SCALE SUMMARIZATION

- Different measures rank summaries differently. Various studies show that intelligent summarizers outperform lead-based summaries.
- Scalable summarization systems were found to be successful using several measures like kappa, relative utility, relevance correlation, and content based for improving accuracy and performance. These measures provide significant advantages over more simple methods like using precision, recall, and percent agreement.
- Applicability to multi-document summaries and the ability to include human and chance agreements is a major challenge in large-scale summarization.
- A new evaluation measure called relevance correlation had been evidently performing better in large-scale summarization systems. This metric estimates how well a summary can be used to replace a document for retrieval purposes.
- The largest and most complete annotated corpus for further research in text summarization had been an open challenge.

CONCLUSION

The various applications in this field are explored, and this chapter gave the scope and applications of text summarization systems in two domains: medical domain and social media domain. This concludes with the recent challenges for large-scale summarization systems.

REFERENCES

Aswin, V. B., Javed, M., Parihar, P., Aswanth, K., Druval, C. R., Dagar, A., and Aravinda, C. V. 2019. "NLP driven ensemble based automatic subtitle generation and semantic video summarization technique." *arXiv preprint arXiv:1904.09740.*

Cheeks, L. H., Gaffar, A., and Moore, M. J. 2017. "Modeling double subjectivity for gaining programmable insights: Framing the case of Uber." *Advances in Science, Technology and Engineering Systems* 2(3): 1677–1692.

Cong, Y., Liu, J., Sun, G., You, Q., Li, Y. and Luo, J. 2017. "Adaptive greedy dictionary selection for web media summarization." *IEEE Transactions on Image Processing,* 26(1): 185–195.

Duan, Y., Chen, Z., Wei, F., Zhou, M., and Shum, H. Y. 2012, December. "Twitter topic summarization by ranking tweets using social influence and content quality." In *Proceedings of COLING 2012Organizing Committee, 2012.* Mumbai (pp. 763–780).

Gupta, B., and Gupta, A. 2019. "Review of various sentiment analysis techniques of Twitter data." *International Journal of Computer Science and Mobile Computing* 8(8): 77–81.

http://www.ncbi.nlm.nih.gov/entrez/query.fcgi

http://www.nlm.nih.gov/mesh/meshhome.html

http://www.promedmail.org

http://www.mdlinx.com/
Maylawati, D. S., Kumar, Y. J., Kasmin, F. B., and Ramdhani, M. A. 2019. "An idea based on sequential pattern mining and deep learning for text summarization." *Journal of Physics: Conference Series* 1402(7): 077013.

Pudaruth, S., Boodhoo, K., and Goolbudun, L. 2016. "An intelligent question answering system for ict." In *2016 International Conference on Electrical, Electronics, and Optimization Techniques (ICEEOT)* (pp. 2895–2899). Chennai, India: IEEE.

Turban, E., Outland, J., King, D., Lee, J. K., Liang, T. P., and Turban, D. C. 2018. "Social commerce: Foundations, social marketing, and advertising." In *Electronic Commerce 2018. Springer Texts in Business and Economics*. Cham, Heidelberg: Springer.

Wang, K., Liu, X., and Han, Y. 2019. "Exploring goodreads reviews for book impact assessment." *Journal of Informetrics* 13(3): 874–886.

Appendix A: Python Projects and Useful Links on Text Summarization

This section will briefly cover the text summarization machine learning algorithms and the implementation aspects in Python with source code. Some of the algorithms and Python libraries available are explained below.

TEXT SUMMARIZATION ALGORITHMS

PageRank Algorithm

Named after Larry Page, one of Google's founders, PageRank is a Google Search algorithm that ranks websites in search engine result pages. Although Google uses multiple algorithms, PageRank is the company's first-ever and most well-known algorithm. PageRank is a means of determining how important a website pages are. PageRank calculates a rough estimate of the importance of a website by tracking the amount and quality of links to that page. The central idea is that more important websites are more likely to gain links from other sites.

The PageRank algorithm generates a probability distribution that indicates the possibility of a random user clicking on links ending up on a specific page. PageRank works for almost any large set of documents. Various research publications imply that the distribution is uniformly distributed among all records in the set at the start of the computational process. PageRank computations demand multiple visits through the collection, also known as 'iterations,' to precisely modify the estimated PageRank values to represent the actual potential value.

TextRank Algorithm

TextRank is an unsupervised extractive text summarizing approach similar to Google's PageRank algorithm. It helps with keyword extraction, automatic text summarization, and phrase ranking.

The TextRank algorithm is quite similar to the PageRank algorithm in many ways, such as:

- TextRank works with sentences, whereas PageRank works with web pages.

- The probability of a web page transition is calculated in the PageRank algorithm, whereas the TextRank algorithm compares the similarity of any two sentences.
- The TextRank approach stores the similarity scores in a square matrix, identical to the M matrix used for the PageRank approach.

SumBasic Algorithm

SumBasic is a multi-document summarization method that finds the frequency distribution of words across all documents. The algorithm prioritizes frequently occurring words in a document over less frequently occurring words in order to provide a precise and accurate summary. It measures the average probability of each sentence based on its word pattern and selects the best-ranking sentence of the most recurring word until the preferred summary length is achieved.

TEXT SUMMARIZATION IN PYTHON

Python has about 137,000 libraries that are useful in domains such as data science, machine learning, data manipulation, and so on. The significance of these libraries stems from the fact that they save you from creating new codes every time the same process runs. Let's look at the most commonly used text summarization Python libraries:

NLTK

NLTK is an acronym for Natural Language Toolkit. It is the most commonly used Python package for handling human language data. It includes libraries for categorization, tokenization, stemming, tagging, parsing, and other text processing tasks. For text summarization, the NLTK employs the TF-IDF approach. The following steps show how NLTK performs text summarization:

- It tokenizes the sentence: this method separates a text into tokens.
- It constructs a frequency matrix, a matrix of all words sentence by sentence. It tracks how many times a word appears in each sentence.
- It generates a matrix by calculating term frequency:

Term frequency (p) = P/D,

where P is the number of times a specific term (term p in the above equation) appears in a document or dataset. D is the total number of terms in the document/data. The term represents a word in a paragraph, while the document/data represents a paragraph.

- A table is formed, including documents based on keywords.
- Generate a matrix by calculating inverse document frequency (IDF):

$\text{IDF}(t) = \log(N/t)$,

where t is the number of times a specific phrase (term t in the equation above) appears in the document/data. N represents the total number of terms in the document/data.

- Multiply the results from both matrices to calculate the TF-IDF and generate a new matrix.
- Using the TF-IDF score, assign weights to the sentences in the text.

Gensim

Gensim is a Python package that relies on other Python libraries such as NumPy and SciPy. Its unique features include:

- Memory independence, which eliminates the need for storing the entire training compilation in RAM at any given time
- Simplified latent semantic analysis and latent Dirichlet allocation
- The ability to create basic similarity queries for documents in their summarization

It is based on corpus theory, vector theory, models, and sparse matrices. Genism implements the TextRank algorithm for text summarization.

Sumy

Sumy is a Python package for summarizing text documents and HTML pages automatically. It employs a variety of techniques and methodologies for summarizing:

1) LEX-RANK

   It's an unsupervised text summarization method based on graph-based sentence centrality scoring. It adopts the technique of looking for comparable sentences, which are likely to be very important. Run pip install lexrank to install it.
2) LUHN

   LUHN summarizes text using a heuristic manner and is based on the recurrence of the most important terms.

SpaCy

Matthew Honnibal and Ines Montani, the creators of the software business Explosion, created SpaCy. It's majorly used for research purposes. CNN models for part-of-speech tagging, dependency parsing, text categorization, and named entity recognition are some of SpaCy's most notable features.

Text summarization using deep learning models

Recurrent neural networks (RNNs), convolutional neural networks (CNNs), and sequence-to-sequence models are the most commonly used abstractive text summarization deep learning models. This section will provide a basic idea of the sequence-to-sequence model, attention mechanism, and transformers (BERT Model).

Sequence-to-Sequence Model (Seq2Seq Model)

The seq2seq framework involves a series of sentences as input and produces another series of sentences as output. In neural machine translation, input sentences are in one language, and output sentences are translated versions of that language. The seq2seq modeling involves two primary strategies: encoder and decoder.

Encoder Model

The encoder model is often used to encode or change the input phrases while also providing feedback at each step. If you use the LSTM layer, this feedback can be an internal state, such as a hidden state or a cell state. Encoder models extract essential information from input texts while keeping the context intact. You will send your input language into the encoder model in neural machine translation, which will record contextual details without affecting the rest of the input sequence. The output sequences are obtained by feeding the encoder model's outputs into the decoder model.

Decoder Mode

The decoder model is used for word-by-word decoding or anticipates the target texts. Decoder input data accepts target sentence input and guesses the next word, which is then passed into the prediction layer below. The words 'start>' (start of target sentence) and 'end>' (end of target sentence) are used by the model to determine what will be the initial variable for predicting the following word and what will be the finishing variable for determining the sentence's ending. You initially give the model the word 'start>' during training, and the model then forecasts the following word: the decoder target data. This word is then used as input data for the next step, resulting in the prediction of the next word.

Attention Mechanism

Before being used for NLP tasks like text summarization, the attention mechanism was used for neural machine translation. A basic encoder-decoder architecture may struggle when providing long sentences as it cannot evaluate long input parts. The attention mechanism helps retain the information that has a considerable impact on the summary. The attention mechanism determines the weight between the output word and each input word at each output word; the weights sum up to 1. The use of weights has the advantage of indicating which input word requires special attention concerning the output word. After processing each input word, the mean value of the decoder's last hidden layers is determined and given to the softmax layer alongside the final hidden layers in the current phase.

Attention mechanism has two categories:

(a) Global attention
(b) Local attention

Global attention: It generates the context vector using all the hidden states from each time step from the encoder model.

Local attention: It generates the context vector using some hidden states from the encoder model in local attention.

Transformers | BERT Model

The Bidirectional Encoder Representations from Transformers (BERT) word embedding method uses a multilayer bidirectional transformer encoder. The transformer neural network uses parallel attention layers instead of sequential recurrence. By merging the representations of the words or phrases, BERT builds a single big transformer. Furthermore, BERT follows an unsupervised approach for pretraining on a significant text volume. Two tokens are added to the text in the BERT model. The initial token Classification (CLS) combines the complete text sequence information. The second token separator (SEP) is used at the end of each sentence. The final text consists of multiple tokens, and each token has three types of embeddings: token, segmentation, and position embeddings.

Here are some of the main reasons behind the popularity of BERT:

- Its key benefit is that it uses bi-directional learning to obtain the context of words from both left to right and right to left at the same time, having been trained on 2.5 billion words.
- Next sentence prediction (NSP) training allows the model to grasp how sentences link to one another during the model training process. As a result, the model can gather more information.
- Since it has been efficiently pretrained on a huge amount of data (English Wikipedia – 2500 words), the BERT model can be used on smaller datasets while still delivering good results.

Some of the models along with their implementation steps are discussed in this section:

Python | Text Summarizer

Today various organizations, be it online shopping, government and private sector organizations, catering and tourism industry, or other institutions that offer customer services, are concerned about their customers and ask for feedback every single time they use their services. Consider the fact that these companies may be receiving enormous amounts of user feedback every single day. And it would become quite tedious for the management to sit and analyze each of those.But, the technologies today have reached an extent where they can do all the tasks of human beings. And the field which makes these things happen is machine learning. Machines have become capable of understanding human languages using natural language processing. Today researches are being done in the field of text analytics. And one such application of text analytics and NLP is a feedback summarizer which helps in summarizing and shortening the text in the user feedback. This can be done by using an algorithm to reduce the bodies of text but keeping its original meaning, or giving a great insight into the original text.

If you're interested in data analytics, you will find learning about natural language processing very useful. Python provides immense library support for NLP. We will be using NLTK – the Natural Language Toolkit – which will serve our purpose right.

Install NLTK module on your system using: sudo pip install nltk

Let's understand the steps:

Step 1: Importing required libraries

There are two NLTK libraries that will be necessary for building an efficient feedback summarizer.

```
from nltk.corpus import stopwords
from nltk.tokenize import word_tokenize, sent_tokenize
```

Terms Used:

- Corpus

  Corpus means a collection of text. It could be datasets of anything containing texts, be it poems by a certain poet, bodies of work by a certain author, etc. In this case, we are going to use a dataset of predetermined stopwords.
- Tokenizers

  It divides a text into a series of tokens. There are three main tokenizers: word, sentence, and regex tokenizer. We will only use the word and sentence tokenizer.

Step 2: Removing stopwords and storing them in a separate array of words.

Stopword

Any word like (is, a, an, the, for) that does not add value to the meaning of a sentence.

For example, let's say we have the sentence

GeeksForGeeks is one of the most useful websites for competitive programing.

After removing stopwords, we can narrow the number of words and preserve the meaning as follows:

['GeeksForGeeks', 'one', 'useful', 'website', 'competitive', 'programming', '.']

Step 3: Create a frequency table of wordsA Python dictionary that'll keep a record of how many times each word appears in the feedback after removing the stopwords. We can use the dictionary over every sentence to know which sentences have the most relevant content in the overall text.

```
stopWords = set(stopwords.words("english"))
words = word_tokenize(text)
freqTable = dict()
```

Step 4: Assign score to each sentence depending on the words it contains and the frequency table.

We can use the sent_tokenize() method to create the array of sentences. Second, we will need a dictionary to keep the score of each sentence, and we will later go through the dictionary to generate the summary.

```
sentences = sent_tokenize(text)
sentenceValue = dict()
```

Step 5: Assign a certain score to compare the sentences within the feedback.A simple approach to compare our scores would be to find the average score of a sentence. The average itself can be a good threshold.

```
sumValues = 0
for sentence in sentenceValue:
    sumValues += sentenceValue[sentence]
average = int(sumValues / len(sentenceValue))
```

Apply the threshold value and store sentences in order in the summary.

CODE: COMPLETE IMPLEMENTATION OF TEXT SUMMARIZER USING PYTHON

```
# importing libraries
import nltk
from nltk.corpus import stopwords
from nltk.tokenize import word_tokenize, sent_tokenize
# Input text - to summarize
text = """ """
```

```
# Tokenizing the text
stopWords = set(stopwords.words("english"))
words = word_tokenize(text)
 # Creating a frequency table to keep the
# score of each word
freqTable = dict()
for word in words:
    word = word.lower()
    if word in stopWords:
       continue
    if word in freqTable:
       freqTable[word] += 1
    else:
       freqTable[word] = 1
# Creating a dictionary to keep the score
# of each sentence
sentences = sent_tokenize(text)
sentenceValue = dict()
for sentence in sentences:
    for word, freq in freqTable.items():
       if word in sentence.lower():
          if sentence in sentenceValue:
              sentenceValue[sentence] += freq
          else:
              sentenceValue[sentence] = freq
 sumValues = 0
for sentence in sentenceValue:
    sumValues += sentenceValue[sentence]
  # Average value of a sentence from the original text
  average = int(sumValues / len(sentenceValue))
  # Storing sentences into our summary.
summary = ''
for sentence in sentences:
    if (sentence in sentenceValue) and
(sentenceValue[sentence] > (1.2 * average)):
       summary += " " + sentence
print(summary)
```

INPUT:

There are many techniques available to generate extractive summarization to keep it simple, I will be using an unsupervised learning approach to find the sentences similarity and rank them. Summarization can be defined as a task of producing a concise and fluent summary while preserving key information and overall meaning. One benefit of this will be, you don't need to train and build a model prior start using it for your project. It's good to understand Cosine similarity to make the best use of the code you are going to see. Cosine similarity is a measure of similarity between two non-zero vectors of an inner product space that measures the cosine of the angle between them. Its measures cosine of the angle between vectors. The angle will be 0 if sentences are similar.

Output

There are many techniques available to generate extractive summarization. Summarization can be defined as a task of producing a concise and fluent summary while preserving key information and overall meaning. One benefit of this will be, you don't need to train and build a model prior start using it for your project. Cosine similarity is a measure of similarity between two non-zero vectors of an inner product space that measures the cosine of the angle between them.

Word count–based text summarizer

```
text = "Enter Text to Summarize"
if text.count(". ") > 20:
    length = int(round(text.count(". ")/10, 0))
else:
    length = 1
nopuch =[char for char in text if char not in string.
punctuation]
nopuch = "".join(nopuch)
processed_text = [word for word in nopuch.split() if word
.lower() not in nltk.corpus.stopwords.words('english')]
word_freq = {}
for word in processed_text:
    if word not in word_freq:
       word_freq[word] = 1
    else:
       word_freq[word] = word_freq[word] + 1
max_freq = max(word_freq.values())
for word in word_freq.keys():
    word_freq[word] = (word_freq[word]/max_freq)
sent_list = nltk.sent_tokenize(text)
sent_score = {}
for sent in sent_list:
    for word in nltk.word_tokenize(sent.lower()):
       if word in word_freq.keys():
          if sent not in sent_score.keys():
             sent_score[sent] = word_freq[word]
          else:
             sent_score[sent] = sent_score[sent] +
word_freq[word]
summary_sents = nlargest(length, sent_score, key=sent_score.get)
summary = " ".join(summary_sents)
print(summary)
```

TEXT SUMMARIZATION USING LSTM

Seq2Seq model is a model that takes a stream of sentences as an input and outputs another stream of sentences. This can be seen in neural machine translation where input sentences are in one language and the output sentences are the translated versions of that language. Encoder and decoder are the two main techniques used in seq2seq modeling. Let's see about them.

ENCODER MODEL:

Encoder model is used to encode or transform the input sentences and generate feedback after every step. This feedback can be an internal state, i.e. hidden state or cell state, if we are using the LSTM layer. Encoder models capture the vital information from the input sentences while maintaining the context throughout.

In neural machine translation, our input language will be passed into the encoder model where it will capture the contextual information without modifying the meaning of the input sequence. Outputs from the encoder model are then passed into the decoder model to get the output sequences.

DECODER MODEL:

The decoder model is used to decode or predict the target sentences word by word. Decoder input data takes the input of target sentences and predicts the next word which is then fed into the next layer for the prediction. '<start>' (start of target sentence) and '<end>' (end of target sentence) are the two words that help the model to know what will be the initial variable to predict the next word and the ending variable to know the ending of the sentence. While training the model, we first provide the word '<start>'; the model then predicts the next word that is the decoder target data. This word is then fed as input data for the next timestep to get the next word prediction.

For example, if our sentence is **'I Love Python'** so we will add '<start>' at the starting and '<end>' at the ending of the sentence; therefore, our sentence will be **'<start> I Love Python <end>';** now let's see how it works.

Timestep	Input data	Target data
1	<start>	I
2	<start> I	Love
3	<start> I Love	Python
4	<start> I Love Python	<end>

As you can see, our input data will start from '<start>' and the target will predict the next word with the help of input data at every timestep. Our input data doesn't contain the last word as our target data at the last timestep is '<end>' which tells us that we have reached the end of our sentence and stop the iteration. In the same way our target data will be one time step ahead as the first word '<start>' is provided by the input data.

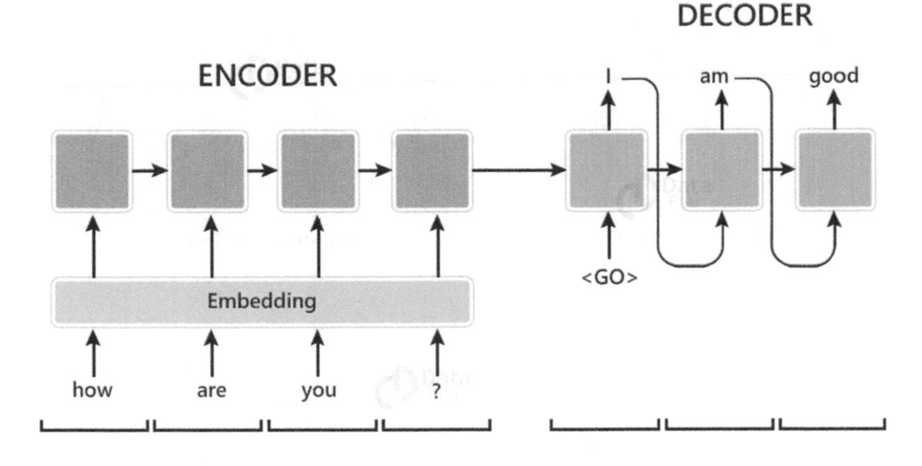

What is an attention mechanism?

Let's take an example to understand the attention mechanism. So below is the input text(review) and target text(summary).

Input text: Now that I've learned about machine learning, I'd like to work on some projects. Can someone recommend the best source for machine learning projects?

Target text: DataFlair is the best source for machine learning projects.

As you can see, we have passed the input text into the model. Rather than focusing on the whole word which is very difficult to remember, we will only focus on specific words for the prediction. In our example, we will only focus on the words like 'source,' 'machine learning,' and 'projects' to predict the target text.

There are two classes of attention mechanisms.

a) Global attention
b) Local attention

Global attention: In global attention, all the hidden states of every time step from the encoder model is used to generate the context vector.

Local attention: In local attention, some of the hidden states from the encoder model is used to generate the context vector.

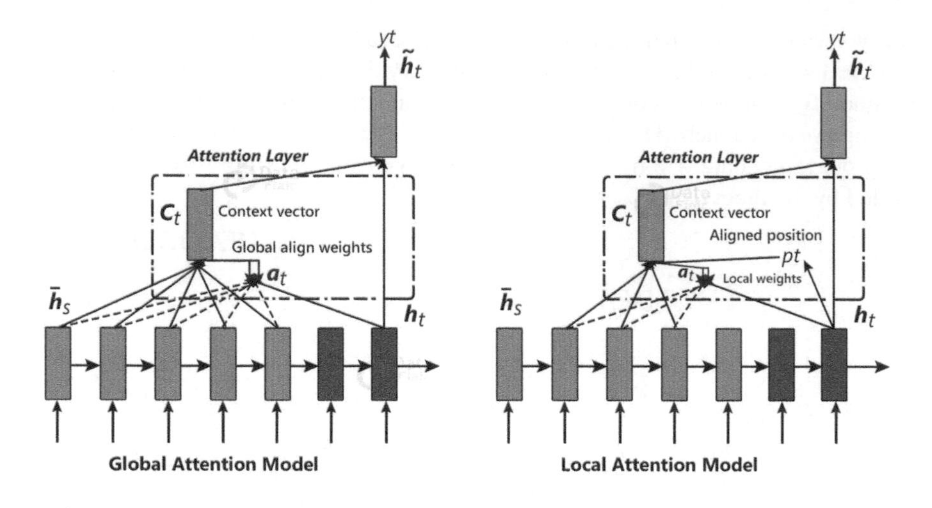

About the project:
In this project, we will use many-to-many sequence models using the abstractive text summarization technique to create models that predict the summary of the reviews. The model will be trained and tested on the first 1,00,000 rows of the dataset file 'Reviews.csv'. Using the attention mechanism, we will focus on specific keywords while maintaining the context of our sentence.

Project prerequisites:
This project requires you to have a good knowledge of Python, deep learning, and natural language processing (NLP). You can install all the modules for this project using the following command:

pip install numpy, pandas, pickle, nltk, tensorflow, sklearn, bs4

The versions which are used in this project for Python and its corresponding modules are as follows:

1) python: 3.8.5
2) TensorFlow: 2.3.1 ***Note*** : tensorFlow version should be 2.2 or higher in order to use Keras or else install Keras directly
3) sklearn: 0.24.2
4) bs4: 4.6.3
5) pickle: 4.0
6) numpy : 1.19.5
7) pandas: 1.1.5
8) nltk : 3.2.5

Text Summarizer Dataset

You can download the dataset file for this project from **Amazon Fine Food Reviews**

Download Text Summarization Project Code

Reviews.csv: This is our dataset file which contains amazon food reviews and summaries.

text_summarizer.py: In this file we will create and train our model with input and target to predict the summary.

s2s/: This directory contains the optimizer, metrics, and weights of our trained model.

contractions.pkl: This file contains a dictionary with key as shortened word and value as its extended or original word.

Steps for Text Summarization:

Import the Libraries

First we will create a file called 'text_summarizer.py' and import all the libraries which have been shared in the prerequisites section.

Code:

```
#DataFlair Project
#import all the required libraries
import numpy as np
import pandas as pd
import pickle
from statistics import mode
import nltk
from nltk import word_tokenize
from nltk.stem import LancasterStemmer
nltk.download('wordnet')
nltk.download('stopwords')
nltk.download('punkt')
from nltk.corpus import stopwords
from tensorflow.keras.models import Model
from tensorflow.keras import models
from tensorflow.keras import backend as K
from tensorflow.keras.preprocessing.sequence import
pad_sequences
from tensorflow.keras.preprocessing.text import Tokenizer
from tensorflow.keras.utils import plot_model
from tensorflow.keras.layers import Input,LSTM,Embedding,
Dense,Concatenate,Attention
from sklearn.model_selection import train_test_split
from bs4 import BeautifulSoup
```

2) Parse the Dataset file.

We will traverse to the dataset file, i.e. ' Reviews.csv' and extract all the input and target texts. For this we will be using the first 1,00,000 rows of our dataset for the training and testing part. It can be changed as per requirements. Our input will be the 'Text' column which is the review column and target will be the 'Summary' column. We will also drop the duplicate records and NA values from our data frame.

Summary	Text
Good Quality Dog Food	I have bought several of the Vitality canned dog food products and have found them all
Not as Advertised	Product arrived labeled as Jumbo Salted Peanuts...the peanuts were actually small size
"Delight" says it all	This is a confection that has been around a few centuries. It is a light, pillowy citrus gela
Cough Medicine	If you are looking for the secret ingredient in Robitussin I believe I have found it. I got thi
Great taffy	Great taffy at a great price. There was a wide assortment of yummy taffy. Delivery was
Nice Taffy	I got a wild hair for taffy and ordered this five pound bag. The taffy was all very enjoyable
Great! Just as good as the expensive br	This saltwater taffy had great flavors and was very soft and chewy. Each candy was ind
Wonderful, tasty taffy	This taffy is so good. It is very soft and chewy. The flavors are amazing. I would definit
Yay Barley	Right now I'm mostly just sprouting this so my cats can eat the grass. They love it. I rotat
Healthy Dog Food	This is a very healthy dog food. Good for their digestion. Also good for small puppies. My
The Best Hot Sauce in the World	I don't know if it's the cactus or the tequila or just the unique combination of ingredients,
My cats LOVE this "diet" food better than	One of my boys needed to lose some weight and the other didn't. I put this food on the
My Cats Are Not Fans of the New Food	My cats have been happily eating Felidae Platinum for more than two years. I just got a
fresh and greasy!	good flavor! these came securely packed... they were fresh and delicious! i love these T
Strawberry Twizzlers - Yummy	The Strawberry Twizzlers are my guilty pleasure - yummy. Six pounds will be around for
Lots of twizzlers, just what you expect.	My daughter loves twizzlers and this shipment of six pounds really hit the spot. It's exact
poor taste	I love eating them and they are good for watching TV and looking at movies! It is not too
Love it!	I am very satisfied with my Twizzler purchase. I shared these with others and we have a
GREAT SWEET CANDY!	Twizzlers, Strawberry my childhood favorite candy, made in Lancaster Pennsylvania by
Home delivered twizlers	Candy was delivered very fast and was purchased at a reasonable price. I was home b

```
#read the dataset file for text Summarizer
df=pd.read_csv("Reviews.csv",nrows=100000)
#drop the duplicate and na values from the records
df.drop_duplicates(subset=['Text'],inplace=True)
df.dropna(axis=0,inplace=True)
input_data = df.loc[:,'Text']
target_data = df.loc[:,'Summary']
target.replace('', np.nan, inplace=True)
```

3) Preprocessing

Real-world texts are incomplete, and they cannot be sent directly to the model, which will cause certain errors. So, we clean all our texts and convert them into a presentable form for prediction tasks. So, first we will initialize all the variables and methods.

```
input_texts=[]
target_texts=[]
input_words=[]
target_words=[]
contractions=pickle.load(open("contractions.pkl","rb"))['co
ntractions']
#initialize stop words and LancasterStemmer
stop_words=set(stopwords.words('english'))
stemm=LancasterStemmer()
```

Some of our texts are in html format and contain html tags, so first we will parse this text and remove all the html tags using 'BeautifulSoup library'. After that, we tokenize our texts into words. And also check the following conditions:

1) Contain integers
2) Are less than three characters
3) Are in stopwords

If one of the above conditions matches, we will remove that particular word from the list of input or target words.

Code:

```
def clean(texts,src):
 #remove the html tags
 texts = BeautifulSoup(texts, "lxml").text
 #tokenize the text into words
 words=word_tokenize(texts.lower())
 #filter words which contains \
 #integers or their length is less than or equal to 3
 words= list(filter(lambda w:(w.isalpha() and len(w)>=3),words))
```

We also have contraction words in our input or target texts that are the combinations of two words, shortened using apostrophe or by dropping letters, for example 'haven't' is shortened for 'have not.' We will expand these kinds of words using the 'contractions.pkl' file which contains a dictionary having keys as shortened words and values as expanded words. Also we will stem all the input words to their root words.

Stemming: Stemming is the process of reducing words into their root words.

For example, if the text contains word like 'chocollate' which might be misspelled for 'chocolate.' If we don't stem our words, then the model will treat them as two different words. Stemmer will stem or reduce that error word to its root word, i.e. 'chocol.' As a result, 'chocol' is the root word for both 'chocolate' and 'chocollate.'

Code:

```
#contraction file to expand shortened words
words= [contractions[w] if w in contractions else w for w in
words ]
#stem the words to their root word and filter stop words
if src=="inputs":
 words= [stemm.stem(w) for w in words if w not in
stop_words]
else:
 words= [w for w in words if w not in stop_words]
return words
We will add 'sos' to the start and 'eos' at the end of
target text to tell our model that this is the starting and
ending of sentences.
```

Code:

```
#pass the input records and taret records
for in_txt,tr_txt in zip(input_data,target_data):
 in_words= clean(in_txt,"inputs")
 input_texts+= [' '.join(in_words)]
 input_words+= in_words
```

```
 #add 'sos' at start and 'eos' at end of text
 tr_words= clean("sos "+tr_txt+" eos","target")
 target_texts+= [' '.join(tr_words)]
 target_words+= tr_words
```

Now after cleaning the sentences, we will filter duplicate words and sort them accordingly. Also we will store the total number of input and target words.

Code:

```
#store only unique words from input and target list of words
input_words = sorted(list(set(input_words)))
target_words = sorted(list(set(target_words)))
num_in_words = len(input_words) #total number of input words
num_tr_words = len(target_words) #total number of target words
 #get the length of the input and target texts which appears
      most often
max_in_len = mode([len(i) for i in input_texts])
max_tr_len = mode([len(i) for i in target_texts])
 print("number of input words : ",num_in_words)
print("number of target words : ",num_tr_words)
print("maximum input length : ",max_in_len)
print("maximum target length : ",max_tr_len)
```

4) Splitting the records

Split the dataset records into training and testing sets. We will be splitting in the 80:20 ratio, where 80% record will be for training sets and 20% for testing sets.

Code:

```
#split the input and target text into 80:20 ratio or testing
size of 20%.
x_train,x_test,y_train,y_test=train_test_split(input_texts,
target_texts,test_size=0.2,random_state=0)
```

5) Text Vectorization

We will convert our word into integer sequence using vectorization technique.
For example,

```
L = [ 'what doing', 'how are you', 'good ']
Tokenize all the elements of list 'L' and make a dictionary
having key as tokens and value as the counter number. So
after the data is fit, we get a dictionary as
D = { 'what' : 1 , 'doing' :2 , 'how' : 3 , 'are' : 4 ,
  'you' :5 , 'good' : 6 }
```

```
So we have fit our data, now let's transform the below list
'J' into integer sequence using our tokenizer.
J = [ 'what are you doing', 'you are good' ]
Transformed (Vectorized) J : [ [ 1 , 4 , 5 , 2 ] , [ 5 , 4 ,
     6 ] ]
```

Code:

```
#train the tokenizer with all the words
in_tokenizer = Tokenizer()
in_tokenizer.fit_on_texts(x_train)
tr_tokenizer = Tokenizer()
tr_tokenizer.fit_on_texts(y_train)
 #convert text into sequence of integers
#where the integer will be the index of that word
x_train= in_tokenizer.texts_to_sequences(x_train)
y_train= tr_tokenizer.texts_to_sequences(y_train)
```

After converting to integer sequence, we will also make all the input and target texts to the same length for our model. So we will take the length of the input sentences which has the highest frequency and store it in the 'max_in_length' variable and repeat the same for target data also. Now we will pad arrays of 0's to the texts if it is less than the assigned maximum input length.

Our encoder input data will be padded 'x_train' and decoder input data will be padded 'y_train,' but we will not include the last word, i.e. 'eos.' Decoder target data will be the same as the decoder input data, but it will be one time step ahead as it will not include the start word, i.e. 'sos' of our target sentence.

Code:

```
#pad array of 0's if the length is less than the maximum
length
en_in_data= pad_sequences(x_train, maxlen=max_in_len,
padding='post')
dec_data= pad_sequences(y_train, maxlen=max_tr_len,
padding='post')
 #decoder input data will not include the last word
#i.e. 'eos' in decoder input data
dec_in_data = dec_data[:,:-1]
#decoder target data will be one time step ahead as it will
not include
# the first word i.e 'sos'
dec_tr_data = dec_data.reshape(len(dec_data),max_tr_len,1)[
    :,1:]
```

6) Build the model.

We are using Stacked LSTM containing three layers of LSTM stacked on top of each other. This will make our prediction much better. As per your requirement, you can have more also. Let's understand our encoder model and decoder model.

Encoder: We will initialize the encoder input tensor using the 'Input' object. The expected shape of the batch will be 74 (maximum input length) dimensions. Then we will create an 'Embedding Layer' which will have the total number of input words as the first argument and a shape of 500 which is the latent (hidden) dimension.

Code:

```
K.clear_session()
latent_dim = 500
 #create input object of total number of encoder words
en_inputs = Input(shape=(max_in_len,))
en_embedding = Embedding(num_in_words+1, latent_dim)
(en_inputs)
```

LSTM: Now we will create three stacked LSTM layers where the first LSTM layer will have input of encoder and like that create a continuous sequence of LSTM layers.

The LSTM layer will capture all the contextual information present in the input sequence. We will return the hidden state output and also states, i.e. hidden state and cell state after execution of every LSTM layer.

Code:

```
#create 3 stacked LSTM layer with the shape of hidden
dimension for text summarizer using deep learning
#LSTM 1
en_lstm1= LSTM(latent_dim, return_state=True,
return_sequences=True)
en_outputs1, state_h1, state_c1= en_lstm1(en_embedding)
 #LSTM2
en_lstm2= LSTM(latent_dim, return_state=True,
return_sequences=True)
en_outputs2, state_h2, state_c2= en_lstm2(en_outputs1)
 #LSTM3
en_lstm3= LSTM(latent_dim,return_sequences=True,return_state
=True)
en_outputs3 , state_h3 , state_c3= en_lstm3(en_outputs2)
#encoder states
en_states= [state_h3, state_c3]
```

Decoder: Like encoder we will initialize the decoder input tensor and then pass it to the only LSTM. Here, the decoder will also have the initial state where we will pass the hidden state and cell state values that we have obtained from the encoder's LSTM layer.

Code:

Decoder.

```
dec_inputs = Input(shape=(None,))
dec_emb_layer = Embedding(num_tr_words+1, latent_dim)
dec_embedding = dec_emb_layer(dec_inputs)
```

```
 #initialize decoder's LSTM layer with the output states of
encoder
dec_lstm = LSTM(latent_dim, return_sequences=True,
return_state=True)
dec_outputs, *_ = dec_lstm(dec_embedding,initial_state=en_
states)
```

Attention layer: We will pass the encoder and decoder outputs into the attention layer, and then we will concatenate attention layer outputs with the decoder outputs.

Code:

```
#Attention layer
attention =Attention()
attn_out = attention([dec_outputs,en_outputs3])
 #Concatenate the attention output with the decoder outputs
merge=Concatenate(axis=-1, name='concat_layer1')([dec_out
puts,attn_out])
```

Now we will create our dense layer, that is, the output layer for our model. It will have the shape of the total number of target words and a softmax activation function.

Code:

```
#Dense layer (output layer)
dec_dense = Dense(num_tr_words+1, activation='softmax')
dec_outputs = dec_dense(merge)
```

5) Train the model.

Finally, we will initialize our model class with the input and output data from the encoder and decoder layers. We can plot the model layers and also get the summary of our model.

Code:

```
#Model class and model summary for text Summarizer
model = Model([en_inputs, dec_inputs], dec_outputs)
model.summary()
plot_model(model, to_file='model_plot.png', show_
shapes=True, show_layer_names=True)
Model Summary and plot:
Model: "model"
_____________________________________________________________
Layer (type) Output Shape Param # Connected to
=============================================================
===================input_5 (InputLayer) [(None, 74)] 0
_____________________________________________________________
embedding (Embedding) (None, 74, 500) 16066000 input_5[0][0]
_____________________________________________________________
```

```
lstm (LSTM) [(None, 74, 500), (N 2002000 embedding[0][0]
____________________________________________________________
input_6 (InputLayer) [(None, None)] 0
____________________________________________________________
lstm_1 (LSTM) [(None, 74, 500), (N 2002000 lstm[0][0]
____________________________________________________________
embedding_1 (Embedding) (None, None, 500) 7079000 input_6[0][0]
____________________________________________________________
lstm_2 (LSTM) [(None, 74, 500), (N 2002000 lstm_1[0][0]
____________________________________________________________
lstm_3 (LSTM) [(None, None, 500), 2002000 embedding_1[0][0]
lstm_2[0][1]
lstm_2[0][2]
____________________________________________________________
attention (Attention) (None, None, 500) 0 lstm_3[0][0]
lstm_2[0][0]
____________________________________________________________
concat_layer1 (Concatenate) (None, None, 1000) 0 lstm_3[0][0]
attention[0][0]
____________________________________________________________
dense (Dense) (None, None, 14158) 14172158 concat_layer1[0][0]
============================================================
===============Total params: 45,325,158
Trainable params: 45,325,158
Non-trainable params: 0
ml model plot
```

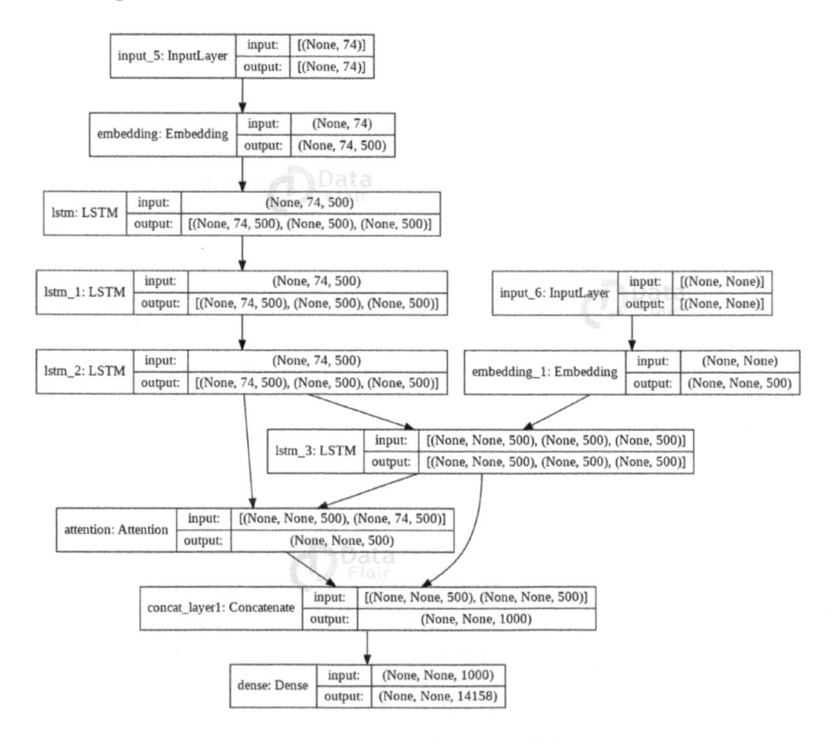

We will pass the data and train our model with '512' batch size, epoch of '10,' and we will be using 'RMSprop' optimizer to train our model. You can increase or decrease the epoch, but take care of the validation loss.

Code:

```
model.compile(
  optimizer="rmsprop", loss="sparse_categorical_
crossentropy", metrics=["accuracy"] )
model.fit(
  [en_in_data, dec_in_data],
  dec_tr_data,
  batch_size=512,
  epochs=10,
  validation_split=0.1,
  )
 #Save model
model.save("s2s")
```

After our model gets trained, we will get a directory as 's2s/' with 'saved _model.pb' which includes optimizer, losses, and metrics of our model. The weights are saved in the variables/directory.

6) Inference Model

We will be using the saved model to create an inference architecture for the encoder and decoder model. The inference model is used to test the new sentences for which the target sequence is not known.

Encoder inference: Input for the inference encoder model will be 0th layer, i.e. input object that we have created (you can check it from the above summary and model plot) and output will be the output of the last LSTM which is the sixth layer.

Code:

```
# encoder inference
latent_dim=500
#/content/gdrive/MyDrive/Text Summarizer/
#load the model
model = models.load_model("s2s")
 #construct encoder model from the output of 6 layer i.e.l
ast LSTM layer
en_outputs,state_h_enc,state_c_enc = model.layers[6].output
en_states=[state_h_enc,state_c_enc]
#add input and state from the layer.
en_model = Model(model.input[0],[en_outputs]+en_states)
```

Decoder inference: Same as the encoder inference model, we will get the input, embedding, and LSTM layers from the saved model. Initialize the decoder hidden input and the other two states with the shape of latent (hidden) dimensions.

Code:

```
# decoder inference
#create Input object for hidden and cell state for decoder
#shape of layer with hidden or latent dimension
dec_state_input_h = Input(shape=(latent_dim,))
dec_state_input_c = Input(shape=(latent_dim,))
dec_hidden_state_input =
Input(shape=(max_in_len,latent_dim))
 # Get the embeddings and input layer from the model
dec_inputs = model.input[1]
dec_emb_layer = model.layers[5]
dec_lstm = model.layers[7]
dec_embedding= dec_emb_layer(dec_inputs)
 #add input and initialize LSTM layer with encoder LSTM
states.
dec_outputs2, state_h2, state_c2 = dec_lstm(dec_embedding,
initial_state=[dec_state_input_h,dec_state_input_c])
```

Attention inference: In our case, the eighth layer is the attention layer. We will fetch it and pass the inference decoder output with the hidden state-input that we have initialized earlier. Then we will concatenate the decoder output with the attention layer output.

Code:

```
#Attention layer
attention = model.layers[8]
attn_out2 = attention([dec_outputs2,dec_hidden_state_input])
 merge2 = Concatenate(axis=-1)([dec_outputs2, attn_out2])
```

And same for the dense layer (output layer) which is the tenth layer of our saved model. Initialize the inference model class with the above data.

Code:

```
#Dense layer
dec_dense = model.layers[10]
dec_outputs2 = dec_dense(merge2)
# Finally define the Model Class
dec_model = Model(
[dec_inputs] + [dec_hidden_state_input,dec_state_input_h,dec
_state_input_c],
[dec_outputs2] + [state_h2, state_c2])
```

Encode the input sequence as state vectors. Create an empty array of the target sequence and generate the start word, i.e. 'sos' in our case, for every pair. Use this state value along with the input sequence to predict the output index. Use the reverse target word index to get the word from the output index and append it to the decoded sequence.

Code:

```
#create a dictionary with a key as index and value as words.
reverse_target_word_index = tr_tokenizer.index_word
reverse_source_word_index = in_tokenizer.index_word
target_word_index = tr_tokenizer.word_index
reverse_target_word_index[0]=' '
def decode_sequence(input_seq):
  #get the encoder output and states by passing the input
       sequence
  en_out, en_h, en_c= en_model.predict(input_seq)
   #target sequence with initial word as 'sos'
  target_seq = np.zeros((1, 1))
  target_seq[0, 0] = target_word_index['sos']
   #if the iteration reaches the end of text than it will
       be stop the iteration
  stop_condition = False
  #append every predicted word in decoded sentence
  decoded_sentence = ""
  while not stop_condition:
    #get predicted output, hidden and cell state.
    output_words, dec_h, dec_c= dec_model.predict([target_
           seq] + [en_out,en_h, en_c])
    #get the index and from the dictionary get the word for
         that index.
    word_index = np.argmax(output_words[0, -1, :])
    text_word = reverse_target_word_index[word_index]
    decoded_sentence += text_word +" "
```

Assign the index of our word to the target sequence, so for the next iteration, our target sequence will have a vector of the previous word. Iterate until our word is equal to the last word, i.e. 'eos' in our case or max length of the target text.

Code:

```
# Exit condition: either hit max length
     # or find a stop word or last word.
     if text_word == "eos" or len(decoded_sentence) >
        max_tr_len:
       stop_condition = True
     #update target sequence to the current word index.
     target_seq = np.zeros((1, 1))
     target_seq[0, 0] = word_index
     en_h, en_c = dec_h, dec_c
 #return the decoded sentence
 return decoded_sentence
```

Finally, we have done all the processes, and now we can predict the summary for the input review.

Code:

```
 inp_review = input("Enter : ")
print("Review :",inp_review)
inp_review = clean(inp_review,"inputs")
inp_review = ' '.join(inp_review)
inp_x= in_tokenizer.texts_to_sequences([inp_review])
inp_x= pad_sequences(inp_x, maxlen=max_in_len,
padding='post')
 summary=decode_sequence(inp_x.reshape(1,max_in_len))
```

if 'eos' in summary:

```
summary=summary.replace('eos','')
print("\nPredicted summary:",summary);print("\n")
Text Summarizer Output
text summarizer output
```

```
Review : I finally received my order for Hot flavor 2.1 lbs. I bought this because of all the
My first thought about the taste of this beef stick is that it tasted really sour and
i tasted some spices but it was far from hot,it wasn't even hot at all. Its easy to chew proba
The reason why i gave it a 2 star instead of 1 star is because it's was going for a good price
I'm sure it's not expired because  it says feb 2013. I knew i should've gotten the
peperoni flavor beef sticks. Also, from all the other people's review i was expecting a sample
I was kinda disappointed. So i won't be buying the hot shots again and its not because i didn'

Predicted summary: worst hot pepper ever


Review : I love the taste of the Republic of Tea teas and I love the fact that they are surgar

Predicted summary: great tea


Review : Product arrived labeled as Jumbo Salted Peanuts...the peanuts were actually small siz
Not sure if this was an error or if the vendor intended to represent the product as Jumbo.

Predicted summary: disappointing
```

Summary

In this project, we have developed a text summarizer model which generates the summary from the provided review using the LSTM model and attention mechanism. We got an accuracy of 87.82% which is good as we have taken only 1,00,000 records for training and testing sets.

Text summary using weighted scores of the words and sentences:

Text summarization is a subdomain of natural language processing (NLP) that deals with extracting summaries from huge chunks of texts. There are two main types of techniques used for text summarization: NLP-based techniques and deep learning–based techniques. In this article, we will see a simple NLP-based technique for text summarization. We will not use any machine learning library in this article. Rather we will simply use Python's NLTK library for summarizing Wikipedia articles.

In [1]:

```
  pip install --upgrade pip
Collecting pip
 Downloading pip-20.1.1-py2.py3-none-any.whl (1.5 MB)
   |████████████████████████████████| 1.5 MB 7.0 MB/s eta 0:00:01
Installing collected packages: pip
 Attempting uninstall: pip
  Found existing installation: pip 20.1
  Uninstalling pip-20.1:
   Successfully uninstalled pip-20.1
Successfully installed pip-20.1.1
```

Note: you may need to restart the kernel to use updated packages.

linkcode

*** Fetching articles from Wikipedia: Before we could summarize Wikipedia articles, we need to fetch them from the web. To do so we will use a couple of libraries. The first library that we need to download is the Beautiful Soup which is very useful Python utility for web scraping. Execute the following command at the command prompt to download the Beautiful Soup utility.****

In [2]:

pip install beautifulsoup4

Requirement already satisfied: beautifulsoup4 in /opt/conda/lib/python3.7/site-packages (4.9.0)

Requirement already satisfied: soupsieve>1.2 in /opt/conda/lib/python3.7/site-packages (from beautifulsoup4) (1.9.4)

Note: you may need to restart the kernel to use updated packages.

In [3]:

pip install lxml

Requirement already satisfied: lxml in /opt/conda/lib/python3.7/site-packages (4.5.0)

Note: you may need to restart the kernel to use updated packages.

**Another important library that we need to parse XML and HTML is the lxml library. Execute the following command at command prompt to download lxml

In [4]:

pip install nltk

Requirement already satisfied: nltk in /opt/conda/lib/python3.7/site-packages (3.2.4)

Requirement already satisfied: six in /opt/conda/lib/python3.7/site-packages (from nltk) (1.14.0)

Note: you may need to restart the kernel to use updated packages.

NLTK is a leading platform for building Python programs to work with human language data. It provides easy-to-use interfaces to over 50 corpora and lexical resources such as WordNet, along with a suite of text processing libraries for classification, tokenization, stemming, tagging, parsing, and semantic reasoning, wrappers for industrial-strength NLP libraries, and an active discussion forum.

Preprocessing

The first preprocessing step is to remove references from the article. Wikipedia references are enclosed in square brackets. The following script removes the square brackets and replaces the resulting multiple spaces by a single space.

Removing Square Brackets and Extra Spaces

The article_text object contains text without brackets. However, we do not want to remove anything else from the article since this is the original article. We will not remove other numbers, punctuation marks, and special characters from this text since we will use this text to create summaries, and weighted word frequencies will be replaced in this article.

To clean the text and calculate the weighted frequencies, we will create another object.

Removing Special Characters and Digits

Now we have two objects article_text, which contains the original article and formatted_article_text which contains the formatted article. We will use formatted_article_text to create weighted frequency histograms for the words and will replace these weighted frequencies with the words in the article_text object.

Converting Text to Sentences

At this point, we have preprocessed the data. Next, we need to tokenize the article into sentences. We will use the article_text object for tokenizing the article to sentence since it contains full stops. The formatted_article_text does not contain any punctuation and therefore cannot be converted into sentences using the full stop as a parameter.

Find the Weighted Frequency of Occurrence

To find the frequency of occurrence of each word, we use the formatted_article_text variable. We used this variable to find the frequency of occurrence since it doesn't contain punctuation, digits, or other special characters.

In the script above, we first store all the English stopwords from the NLTK library into a stopword variable. Next, we loop through all the sentences and then the corresponding words to first check if they are stopwords. If not, we proceed to check whether the words exist in word_frequency dictionary, i.e.

word_frequencies, or not. If the word is encountered for the first time, it is added to the dictionary as a key and its value is set to 1. Otherwise, if the word previously exists in the dictionary, its value is simply updated by 1.

Finally, to find the weighted frequency, we can simply divide the number of occurrences of all the words by the frequency of the most occurring word.

Calculating Sentence Scores

We have now calculated the weighted frequencies for all the words. Now is the time to calculate the scores for each sentence by adding weighted frequencies of the words that occur in that particular sentence.

In the script above, we first create an empty sentence_scores dictionary. The keys of this dictionary will be the sentences themselves, and the values will be the corresponding scores of the sentences. Next, we loop through each sentence in the sentence_list and tokenize the sentence into words.

We then check if the word exists in the word_frequencies dictionary. This check is performed since we created the sentence_list list from the article_text object; on the other hand, the word frequencies were calculated using the formatted_article_text object, which doesn't contain any stopwords, numbers, etc.

We do not want very long sentences in the summary; therefore, we calculate the score for only sentences with less than 30 words (although you can tweak this parameter for your own use-case). Next, we check whether the sentence exists in the sentence_scores dictionary or not. If the sentence doesn't exist, we add it to the sentence_scores dictionary as a key and assign it the weighted frequency of the first word in the sentence, as its value. On the contrary, if the sentence exists in the dictionary, we simply add the weighted frequency of the word to the existing value.

Getting the Summary

Now we have the sentence_scores dictionary that contains sentences with their corresponding score. To summarize the article, we can take top N sentences with the highest scores. The following script retrieves top 7 sentences and prints them on the screen.

In the script above, we use the heapq library and call its nlargest function to retrieve the top 7 sentences with the highest scores.

In [5]:

```
import bs4 as bs
import urllib.request
import re
import nltk
scraped_data = urllib.request.urlopen('https://en.wikipedia
.org/wiki/Severe_acute_respiratory_syndrome_coronavirus_2')
article = scraped_data.read()
parsed_article = bs.BeautifulSoup(article,'lxml')
paragraphs = parsed_article.find_all('p')
article_text = ""
```

```
for p in paragraphs:
    article_text += p.text
# Removing Square Brackets and Extra Spaces
article_text = re.sub(r'\[[0-9]*\]', ' ', article_text)
article_text = re.sub(r'\s+', ' ', article_text)
# Removing special characters and digits
formatted_article_text = re.sub('[^a-zA-Z]', ' ', article_
text )
formatted_article_text = re.sub(r'\s+', ' ',
formatted_article_text)
sentence_list = nltk.sent_tokenize(article_text)
stopwords = nltk.corpus.stopwords.words('english')
word_frequencies = {}
for word in nltk.word_tokenize(formatted_article_text):
    if word not in stopwords:
       if word not in word_frequencies.keys():
          word_frequencies[word] = 1
       else:
          word_frequencies[word] += 1
    maximum_frequncy = max(word_frequencies.values())
for word in word_frequencies.keys():
    word_frequencies[word] = (word_frequencies[word]/maximu
m_frequncy)
    sentence_scores = {}
for sent in sentence_list:
    for word in nltk.word_tokenize(sent.lower()):
       if word in word_frequencies.keys():
          if len(sent.split(' ')) < 30:
              if sent not in sentence_scores.keys():
                 sentence_scores[sent] = word_frequencies[word]
              else:
                 sentence_scores[sent] += word_frequencies[word]
import heapq
summary_sentences = heapq.nlargest(7, sentence_scores, key
=sentence_scores.get)
summary = ' '.join(summary_sentences)
print(summary)
```

Severe acute respiratory syndrome coronavirus 2 (SARS-CoV-2) is the strain of coronavirus that causes coronavirus disease 2019 (COVID-19), a respiratory illness. During the initial outbreak in Wuhan, China, the virus was commonly referred to as 'coronavirus' or 'Wuhan coronavirus.' In March 2020, US President Donald Trump referred to the virus as the 'Chinese virus' in tweets, interviews, and White House press briefings. Based on whole genome sequence similarity, a pangolin coronavirus candidate strain was found to be less similar than RaTG13, but more similar than other bat coronaviruses to SARS-CoV-2. Arinjay Banerjee, a virologist at McMaster University, notes that 'the SARS virus shared 99.8% of its genome with a civet coronavirus, which is why civets were considered the source.' The virion then releases RNA into the cell and forces the cell to produce and disseminate copies of the virus, which infect more cells. The general public often call 'coronavirus' both the virus and the disease it causes.

Step 1: Installing text summarization Python environment

To follow along with the code in this article, you can download and install our pre-built Text Summarization environment, which contains a version of Python 3.8 and the packages used in this post.

In order to download this ready-to-use Python environment, you will need to create an ActiveState Platform account. Just use your GitHub credentials or your email address to register. Signing up is easy, and it unlocks the many benefits of the ActiveState Platform!

For Linux users: run the following to automatically download and install our CLI, the State Tool, along with the Text Summarization into a virtual environment:

sh <(curl -q https://platform.activestate.com/dl/cli/install.sh) --activate-default Pizza-Team/Text-Summarization

Step 2: Choose a text source for abstractive text summarization

The quality, type, and density of information conveyed via text vary from source to source. Textbooks tend to be low in density but high in quality, while academic articles are high in both quality and density. On the other hand, news articles can vary significantly from source to source.

Regardless of where the text comes from, the goal here is to minimize the time you spend reading. Thus, we will build a tool that can easily be adapted to any number of sources.

For this example, we will use a news article on a recent global warming study from *ScienceDaily* as our text source. Feel free to use a different article.

To extract the text from the URL, we'll use the newspaper3k package:

from newspaper import Article

```
url = 'https://www.sciencedaily.com/releases/2021/08
/210811162816.htm'
article = Article(url)
article.download()
article.parse()
```

Now, we'll download and parse the article to extract the relevant attributes. From here, we can view the article text:

article.text

```
'It is increasingly clear that the prolonged drought conditions, record-breaking heat, sustained wildfires, and frequ
ent, more extreme storms experienced in recent years are a direct result of rising global temperatures brought on by
humans\' addition of carbon dioxide to the atmosphere. And a new MIT study on extreme climate events in Earth\'s anci
ent history suggests that today\'s planet may become more volatile as it continues to warm.\n\nThe study, appearing t
oday in Science Advances, examines the paleoclimate record of the last 66 million years, during the Cenozoic era, whi
ch began shortly after the extinction of the dinosaurs. The scientists found that during this period, fluctuations in
the Earth\'s climate experienced a surprising "warming bias." In other words, there were far more warming events -- p
eriods of prolonged global warming, lasting thousands to tens of thousands of years -- than cooling events. What\'s m
ore, warming events tended to be more extreme, with greater shifts in temperature, than cooling events.\n\nThe resear
chers say a possible explanation for this warming bias may lie in a "multiplier effect," whereby a modest degree of w
arming -- for instance from volcanoes releasing carbon dioxide into the atmosphere -- naturally speeds up certain bio
logical and chemical processes that enhance these fluctuations, leading, on average, to still more warming.\n\nIntere
stingly, the team observed that this warming bias disappeared about 5 million years ago, around the time when ice she
ets started forming in the Northern Hemisphere. It\'s unclear what effect the ice has had on the Earth\'s response to
climate shifts. But as today\'s Arctic ice recedes, the new study suggests that a multiplier effect may kick back in,
and the result may be a further amplification of human-induced global warming.\n\n"The Northern Hemisphere\'s ice she
ets are shrinking, and could potentially disappear as a long-term consequence of human actions" says the study\'s lea
d author Constantin Arnscheidt, a graduate student in MIT\'s Department of Earth, Atmospheric and Planetary Sciences.
"Our research suggests that this may make the Earth\'s climate fundamentally more susceptible to extreme, long-term g
lobal warming events such as those seen in the geologic past."\n\nArnscheidt\'s study co-author is Daniel Rothman, pr
ofessor of geophysics at MIT, and co-founder and co-director of MIT\'s Lorenz Center.\n\nadvertisement\n\nA volatile
push\n\nFor their analysis, the team consulted large databases of sediments containing deep-sea benthic foraminifera
-- single-celled organisms that have been around for hundreds of millions of years and whose hard shells are preserve
d in sediments. The composition of these shells is affected by the ocean temperatures as organisms are growing; the s
hells are therefore considered a reliable proxy for the Earth\'s ancient temperatures.\n\nFor decades, scientists hav
e analyzed the composition of these shells, collected from all over the world and dated to various time periods, to t
rack how the Earth\'s temperature has fluctuated over millions of years.\n\n"When using these data to study extreme c
limate events, most studies have focused on individual large spikes in temperature, typically of a few degrees Celsiu
s warming," Arnscheidt says. "Instead, we tried to look at the overall statistics and consider all the fluctuations i
nvolved, rather than picking out the big ones."\n\nThe team first carried out a statistical analysis of the data and
observed that, over the last 66 million years, the distribution of global temperature fluctuations didn\'t resemble a
standard bell curve, with symmetric tails representing an equal probability of extreme warm and extreme cool fluctuat
ions. Instead, the curve was noticeably lopsided, skewed toward more warm than cool events. The curve also exhibited
a noticeably longer tail, representing warm events that were more extreme, or of higher temperature, than the most ex
treme cold events.\n\nadvertisement\n\n"This indicates there\'s some sort of amplification relative to what you would
otherwise have expected," Arnscheidt says. "Everything\'s pointing to something fundamental that\'s causing this pus
h, or bias toward warming events."\n\n"It\'s fair to say that the Earth system becomes more volatile, in a warming se
nse," Rothman adds.\n\nA warming multiplier\n\nThe team wondered whether this warming bias might have been a result o
f "multiplicative noise" in the climate-carbon cycle. Scientists have long understood that higher temperatures, up to
a point, tend to speed up biological and chemical processes. Because the carbon cycle, which is a key driver of long-
term climate fluctuations, is itself composed of such processes, increases in temperature may lead to larger fluctuat
ions, biasing the system towards extreme warming events.\n\nIn mathematics, there exists a set of equations that desc
ribes such general amplifying, or multiplicative effects. The researchers applied this multiplicative theory to their
analysis to see whether the equations could predict the asymmetrical distribution, including the degree of its skew a
nd the length of its tails.\n\nIn the end, they found that the data, and the observed bias toward warming, could be e
xplained by the multiplicative theory. In other words, it\'s very likely that, over the last 66 million years, period
s of modest warming were on average further enhanced by multiplier effects, such as the response of biological and ch
emical processes that further warmed the planet.\n\nAs part of the study, the researchers also looked at the correlat
ion between past warming events and changes in Earth\'s orbit. Over hundreds of thousands of years, Earth\'s orbit ar
ound the sun regularly becomes more or less elliptical. But scientists have wondered why many past warming events app
eared to coincide with these changes, and why these events feature outsized warming compared with what the change in
Earth\'s orbit could have wrought on its own.\n\nSo, Arnscheidt and Rothman incorporated the Earth\'s orbital changes
into the multiplicative model and their analysis of Earth\'s temperature changes, and found that multiplier effects c
ould predictably amplify, on average, the modest temperature rises due to changes in Earth\'s orbit.\n\n"Climate warm
s and cools in synchrony with orbital changes, but the orbital cycles themselves would predict only modest changes in
climate," Rothman says. "But if we consider a multiplicative model, then modest warming, paired with this multiplier
effect, can result in extreme events that tend to occur at the same time as these orbital changes."\n\n"Humans are fo
rcing the system in a new way," Arnscheidt adds. "And this study is showing that, when we increase temperature, we\'r
e likely going to interact with these natural, amplifying effects."\n\nThis research was supported, in part, by MIT
\'s School of Science.'
```

Clearly, this is quite long and dense. This text will serve as our input for the summarization algorithm that we'll write in the next step.

If your particular application requires extracting text from pdf documents, try out the PyPDF2 package. Alternatively, if you have audio files that need to be transcribed to text, try using the SpeechRecognition package. Once you have the text in a format that Python can understand, you can move on to summarizing it.

Step 3: Summarizing text with SpaCy

A human might approach the task of summarizing a document as follows:

Read the full text

Understand the concepts being conveyed

Pick out the most important concepts

Simplify them in a more concise manner

For a computer to perform the same task, a semantic understanding of the text is necessary. While semantic analysis is possible with current NLP algorithms, it often requires significant computational power and produces results similar to other extractive techniques in quality.

Rather than understanding the text, extractive summarization relies on quantitative metrics constructed from the text itself, without attaching any exogenous meaning. Our approach is to simply:

Look at the use frequency of specific words

Sum the frequencies within each sentence

Rank the sentences based on this sum

Of course, our assumption is that a higher-frequency word use implies a more 'significant' meaning. This may seem overly simplistic, but this approach often produces surprisingly good results.

To begin, we'll first need to import the different packages:

```
import spacy
from spacy.lang.en.stop_words import STOP_WORDS
from string import punctuation
from heapq import nlargest
```

We'll use SpaCy to import a pretrained NLP pipeline to help interpret the grammatical structure of the text. This will allow us to identify the most common words that are often useful to filter out (i.e. STOP_WORDS) as well as the punctuation (i.e. punctuation). We'll also use the nlargest function to extract a percentage of the most important sentences. Our algorithm will use the following steps:

Tokenize the text with the SpaCy pipeline. This segments the text into words, punctuation, and so on, using grammatical rules specific to the English language.

Count the number of times a word is used (not including stopwords or punctuation), then normalize the count. A word that's used more frequently has a higher normalized count.

Calculate the sum of the normalized count for each sentence.

Extract a percentage of the highest-ranked sentences. These serve as our summary.

We can write a function that performs these steps as follows:

```
def summarize(text, per):
    nlp = spacy.load('en_core_web_sm')
    doc= nlp(text)
    tokens=[token.text for token in doc]
    word_frequencies={}
    for word in doc:
        if word.text.lower() not in list(STOP_WORDS):
            if word.text.lower() not in punctuation:
                if word.text not in word_frequencies.keys():
```

```
                word_frequencies[word.text] = 1
            else:
                word_frequencies[word.text] += 1
    max_frequency=max(word_frequencies.values())
    for word in word_frequencies.keys():
        word_frequencies[word]=word_frequencies[word]/max_
frequency
    sentence_tokens= [sent for sent in doc.sents]
    sentence_scores = {}
    for sent in sentence_tokens:
        for word in sent:
            if word.text.lower() in word_frequencies.keys():
                if sent not in sentence_scores.keys():
                    sentence_scores[sent]=word_frequencies[word.
                        text.lower()]
                else:
                    sentence_scores[sent]+=word_frequencies[word
                        .text.lower()]
    select_length=int(len(sentence_tokens)*per)
    summary=nlargest(select_length, sentence_scores,key
                =sentence_scores.get)
    final_summary=[word.text for word in summary]
    summary=''.join(final_summary)
    return summary
```

Note that per is the percentage (0–1) of sentences you want to extract. To test it out on the *ScienceDaily* article, run:

summarize(article.text, 0.05)

The output should look like this:

```
'\n\nThe researchers say a possible explanation for this warming bias may lie in a "multiplier effect," whereby a mod
est degree of warming -- for instance from volcanoes releasing carbon dioxide into the atmosphere -- naturally speeds
up certain biological and chemical processes that enhance these fluctuations, leading, on average, to still more warm
ing." In other words, there were far more warming events -- periods of prolonged global warming, lasting thousands to
tens of thousands of years -- than cooling events.'
```

You can read the complete article to judge how well this reflects the complete text. However, a summary is already provided by the author at the top of the article. It reads, 'Global warming begets more, extreme warming, new paleoclimate study finds. Researchers observe a "warming bias" over the past 66 million years that may return if ice sheets disappear.'

Pretty much spot on, right?

Conclusion: Use text summarization and improve your productivity with Python

Maximizing your efficiency by minimizing the time you spend reading can have a dramatic impact on productivity. Whether you're reading textbooks, reports, or academic journals, the power of natural language processing with Python and SpaCy can reduce the time you spend without diluting the quality of information.

Next steps:

Text summarization steps

- Obtain data
- Text preprocessing
- Convert paragraphs to sentences
- Tokenizing the sentences
- Find the weighted frequency of occurrence
- Replace words by weighted frequency in sentences
- Sort sentences in descending order of weights
- Summarizing the article
- Obtain data for summarization

If you wish to summarize a Wikipedia article, obtain the URL for the article you wish to summarize. We will obtain data from the URL using the concept of Web scraping. Now, to use Web scraping, you will need to install the Beautiful Soup library in Python. This library will be used to fetch the data on the web page within the various HTML tags.

Use the below command:

```
  pip install beautifulsoup4
To parse the HTML tags we will further require a parser,
that is the lxml package:
pip install lxml
```

```
We will try to summarize the Reinforcement Learning page on
Wikipedia.Python Code for obtaining the data through web
scraping:
import bs4 as bs
import urllib.request
import re
scraped_data = urllib.request.urlopen('https://en.wikipedia
.org/wiki/Reinforcement_learning')
article = scraped_data.read()
parsed_article = bs.BeautifulSoup(article,'lxml')
paragraphs = parsed_article.find_all('p')
article_text = ""
for p in paragraphs:
    article_text += p.text
```

In this script, we first begin with importing the required libraries for web scraping, i.e. Beautiful Soup. The urllib package is required for parsing the URL. Re is the library for regular expressions that are used for text pre-processing. The urlopen function will be used to scrape the data. The read() will read the data on the URL. Further on, we will parse the data with the help of the Beautiful Soup object and the lxml parser.

In Wikipedia articles, the text is present in the <p> tags. Hence we are using the find_all function to retrieve all the text which is wrapped within the <p> tags.

After scraping, we need to perform data preprocessing on the text extracted.

Text Preprocessing

The first task is to remove all the references made in the Wikipedia article. These references are all enclosed in square brackets. The below code will remove the square brackets and replace them with spaces.

```
# Removing Square Brackets and Extra Spaces
article_text = re.sub(r'[[0-9]*]', ' ', article_text)
article_text = re.sub(r's+', ' ', article_text)
```

The article_text will contain text without brackets which is the original text. We are not removing any other words or punctuation marks as we will use them directly to create the summaries.

Execute the below code to create the weighted frequencies and also to clean the text:

```
# Removing special characters and digits
formatted_article_text = re.sub('[^a-zA-Z]', ' ', article_
text )
formatted_article_text = re.sub(r's+', ' ',
formatted_article_text)
```

Here the formatted_article_text contains the formatted article. We will use this object to calculate the weighted frequencies, and we will replace the weighted frequencies with words in the article_text object.

Convert text to sentences

The sentences are broken down into words, so that we have separate entities.

sentence_list = nltk.sent_tokenize(article_text)

We are tokenizing the article_text object as it is unfiltered data while the formatted_article_text object has formatted data devoid of punctuations, etc.

Finding weighted frequencies of occurrence

```
stopwords = nltk.corpus.stopwords.words('english')
word_frequencies = {}
for word in nltk.word_tokenize(formatted_article_text):
    if word not in stopwords:
       if word not in word_frequencies.keys():
          word_frequencies[word] = 1
       else:
          word_frequencies[word] += 1
```

All English stopwords from the NLTK library are stored in the stopwords variable. Iterate over all the sentences, and check if the word is a stopword. If the word is not a stopword, then check for its presence in the word_frequencies dictionary. If it doesn't exist, then insert it as a key and set its value to 1. If it is already existing, just increase its count by 1.

```
maximum_frequncy = max(word_frequencies.values())
for word in word_frequencies.keys():
    word_frequencies[word] = (word_frequencies[word]/maximu
m_frequncy)
To find the weighted frequency, divide the frequency of the
word by the frequency of the most occurring word.
A glimpse of the word_frequencies dictionary:
{'Reinforcement': 0.06944444444444445,
 'learning': 0.4583333333333333,
 'RL': 0.013888888888888888,
 'area': 0.013888888888888888,
 'machine': 0.041666666666666664,
 'concerned': 0.027777777777777776,
 'software': 0.013888888888888888,
 'agents': 0.013888888888888888,
 'ought': 0.013888888888888888,
 'take': 0.027777777777777776,
 'actions': 0.1527777777777778,
 'environment': 0.08333333333333333,
 'order': 0.041666666666666664,
```

```
 'maximize': 0.041666666666666664,
 'notion': 0.027777777777777776,
 'cumulative': 0.041666666666666664,
...........
```

Calculate sentence scores

We have calculated the weighted frequencies. Now scores for each sentence can be calculated by adding weighted frequencies for each word.

```
sentence_scores = {}
for sent in sentence_list:
     for word in nltk.word_tokenize(sent.lower()):
         if word in word_frequencies.keys():
             if len(sent.split(' ')) < 30:
                 if sent not in sentence_scores.keys():
                     sentence_scores[sent] = word_frequencies[word]
                 else:
                     sentence_scores[sent] += word_frequencies[word]
```

The sentence_scores dictionary has been created which will store the sentences as keys and their occurrence as values. Iterate over all the sentences, tokenize all the words in a sentence. If the word exists in word_frequences and also if the sentence exists in sentence_scores, then increase its count by 1, else insert it as a key in the sentence_scores and set its value to 1. We are not considering longer sentences; hence, we have set the sentence length to 30.

A glimpse of sentence_scores dictionary:

{'Reinforcement learning is one of three basic machine learning paradigms, alongside supervised learning and unsupervised learning.': 2.3472222222222222,
'Reinforcement learning differs from supervised learning in not needing labeled input/output pairs to be presented, and in not needing sub-optimal actions to be explicitly corrected.': 1.5555555555555551,
'Instead, the focus is on finding a balance between exploration (of uncharted territory) and exploitation (of current knowledge).': 0.4305555555555556,

Summary of the Article

The sentence_scores dictionary consists of the sentences along with their scores. Now, top N sentences can be used to form the summary of the article.Here the heapq library has been used to pick the top 7 sentences to summarize the article.

```
import heapq
summary_sentences = heapq.nlargest(7, sentence_scores, key
=sentence_scores.get)
```

```
summary = ' '.join(summary_sentences)
print(summary)
```

Output:

Reinforcement learning is one of three basic machine learning paradigms, alongside supervised learning and unsupervised learning. In reinforcement learning methods, expectations are approximated by averaging over samples and using function approximation techniques to cope with the need to represent value functions over large state-action spaces. Policy iteration consists of two steps: policy evaluation and policy improvement. The work on learning ATARI games by Google DeepMind increased attention to deep reinforcement learning or end-to-end reinforcement learning. Assuming full knowledge of the MDP, the two basic approaches to compute the optimal action-value function are value iteration and policy iteration. Two elements make reinforcement learning powerful: the use of samples to optimize performance and the use of function approximation to deal with large environments. Many policy search methods may get stuck in local optima (as they are based on local search).

Endnotes

Text summarization of articles can be performed by using the NLTK library and the Beautiful Soup library. This can help in saving time. Higher deep learning techniques can be further used to get more optimum summarizations. Looking forward to people using this mechanism for summarization.

Extractive Summarization Using NLTK Corpus

Steps for implementation

Step 1: The first step is to import the required libraries. There are two NLTK libraries that are necessary for building an efficient text summarizer.

1 from nltk.corpus import stopwords

2 from nltk.tokenize import word_tokenize, sent_tokenize

Terms Used:

Corpus

A collection of text is known as corpus. This could be datasets such as bodies of work by an author, poems by a particular poet, etc. To explain this concept in the blog, we will be using a dataset of predetermined stopwords.

Tokenizers

This divides a text into a series of tokens. In tokenizers, there are three main tokens – sentence, word, and regex tokenizer. We will be using only the word and the sentence tokenizer.

Step 2: Remove the stopwords and store them in a separate array of words.

Stopwords

Words such as **is, an, a, the, for** that do not add value to the meaning of a sentence. For example, let us take a look at the following sentence:

GreatLearning is one of the most useful websites for ArtificialIntelligence aspirants.

After removing the stopwords in the above sentence, we can narrow down the number of words and preserve the meaning as follows:

['GreatLearning', 'one', 'useful', 'website', 'ArtificialIntelligence', 'aspirants', '.']

Step 3: We can then create a frequency table of the words.

A Python dictionary can keep a record of how many times each word will appear in the text after removing the stopwords. We can use this dictionary over each sentence to know which sentences have the most relevant content in the overall text.

```
1 stopwords = set (stopwords.words("english"))
2 words = word_tokenize(text)
3 freqTable = dict()
```

Step 4: Depending on the words it contains and the frequency table, we will assign a score to each sentence.

Here, we will use the **sent_tokenize()** method that can be used to create the array of sentences. We will also need a dictionary to keep track of the score of each sentence, and we can later go through the dictionary to create a summary.

```
1 sentences = sent_tokenize(text)
2 sentenceValue = dict()
```

Step 5: To compare the sentences within the text, assign a score.

One simple approach that can be used to compare the scores is to find an average score of a particular sentence. This average score can be a good threshold.

```
1 sumValues = 0
2 for sentence in sentenceValue:
3 sumValues += sentenceValue[sentence]
4 average = int(sumValues / len(sentenceValue))
```

Apply the threshold value and the store sentences in an order into the summary.

Complete Code:

```
# importing libraries
import nltk
from nltk.corpus import stopwords
from nltk.tokenize import word_tokenize, sent_tokenize

# Input text - to summarize
text = """ There are many techniques available to generate extractive summarization to keep it simple,
I will be using an unsupervised learning approach to find the sentences similarity and rank them.
Summarization can be defined as a task of producing a concise and fluent summary while preserving key
information and overall meaning. One benefit of this will be, you don't need to train and build a model
prior start using it for your project. It's good to understand Cosine similarity to make the best use of
the code you are going to see. Cosine similarity is a measure of similarity between two non-zero vectors
of an inner product space that measures the cosine of the angle between them. Its measures cosine of the
angle between vectors. The angle will be 0 if sentences are similar"""

# Tokenizing the text
stopWords = set(stopwords.words("english"))
words = word_tokenize(text)

# Creating a frequency table to keep the
# score of each word

freqTable = dict()
for word in words:
    word = word.lower()
    if word in stopWords:
        continue
    if word in freqTable:
        freqTable[word] += 1
    else:
        freqTable[word] = 1

# Creating a dictionary to keep the score
# of each sentence
sentences = sent_tokenize(text)
sentenceValue = dict()

for sentence in sentences:
    for word, freq in freqTable.items():
        if word in sentence.lower():
            if sentence in sentenceValue:
                sentenceValue[sentence] += freq
            else:
                sentenceValue[sentence] = freq


sumValues = 0
for sentence in sentenceValue:
    sumValues += sentenceValue[sentence]

# Average value of a sentence from the original text

average = int(sumValues / len(sentenceValue))

# Storing sentences into our summary.
summary = ''
for sentence in sentences:
    if (sentence in sentenceValue) and (sentenceValue[sentence] > (1.2 * average)):
        summary += " " + sentence
print(summary)
```

Input Text

There are many techniques available to generate extractive summarization to keep it simple, I will be using an unsupervised learning approach to find the sentences similarity and rank the m. Summarization can be defined as a task of producing a concise and fluent summary while pr eserving key information and overall meaning. One benefit of this will be, you don't need to trai n and build a model prior start using it for your project. It's good to understand Cosine similarit y to make the best use of the code you are going to see. Cosine similarity is a measure of similar ity between two non-zero vectors of an inner product space that measures the cosine of the an gle between them. Its measures cosine of the angle between vectors. The angle will be 0 if sen tences are similar.

Output Summary

> *There are many techniques available to generate extractive summarization. Summarization ca n be defined as a task of producing a concise and fluent summary while preserving key informa tion and overall meaning. One benefit of this will be, you don't need to train and build a model p rior start using it for your project. Cosine similarity is a measure of similarity between two non-zero vectors of an inner product space that measures the cosine of the angle between them.*

1. Import all necessary libraries
from nltk.corpus import stopwords
from nltk.cluster.util import cosine_distance
import numpy as np
import networkx as nx
2. Generate clean sentences
def read_article(file_name):

```
    file = open(file_name, "r")
    filedata = file.readlines()
    article = filedata[0].split(". ")
    sentences = []  for sentence in article:
     print(sentence)
     sentences.append(sentence.replace("[^a-zA-Z]", "
").split(" "))
     sentences.pop()
       return sentences
```

3. Similarity matrix

This is where we will be using cosine similarity to find the similarity between sentences.

def build_similarity_matrix(sentences, stop_words):

```
    # Create an empty similarity matrix
    similarity_matrix = np.zeros((len(sentences),
len(sentences)))
    for idx1 in range(len(sentences)):
       for idx2 in range(len(sentences)):
          if idx1 == idx2: #ignore if both are same sentences
             continue
          similarity_matrix[idx1][idx2] = sentence_
            similarity(sentences[idx1], sentences[idx2],
            stop_words)return similarity_matrix
```

4. Generate summary method

The method will keep calling all other helper functions to keep our summarization pipeline going. Make sure to take a look at all # Steps in the below code.

def generate_summary(file_name, top_n=5):

```
stop_words = stopwords.words('english')
summarize_text = []    # Step 1 - Read text and tokenize
sentences = read_article(file_name) # Step 2 - Generate
  Similary Martix across sentences  sentence_similarity_
  martix = build_similarity_matrix(sentences, stop_
  words) # Step 3 - Rank sentences in similarity matrix
sentence_similarity_graph = nx.from_numpy_array(sente
  nce_similarity_martix)
scores = nx.pagerank(sentence_similarity_graph)  # Step
  4 - Sort the rank and pick top sentences
ranked_sentence = sorted(((scores[i],s) for i,s in
  enumerate(sentences)), reverse=True)
print("Indexes of top ranked_sentence order are ",
  ranked_sentence)for i in range(top_n):
 summarize_text.append(" ".join(ranked_sentence[i][1]))
  # Step 5 - Offcourse, output the summarize texr
print("Summarize Text: \n", ". ".join(summarize_text))
```

All put together, here is the complete code.

The complete text from an article titled *Microsoft Launches Intelligent Cloud Hub to Upskill Students in AI & Cloud Technologies*

In an attempt to build an AI-ready workforce, Microsoft announced Intelligent Cloud Hub which has been launched to empower the next generation of students with AI-ready skills. Envisioned as a three-year collaborative program, Intelligent Cloud Hub will support around 100 institutions with AI infrastructure, course content and curriculum, developer support, development tools and give students access to cloud and AI services. As part of the program, the Redmond giant which wants to expand its reach and is planning to build a strong developer ecosystem in India with the program will set up the core AI infrastructure and IoT Hub for the selected campuses. The company will provide AI development tools and Azure AI services such as Microsoft Cognitive Services, Bot Services and Azure Machine Learning. According to Manish Prakash, Country General Manager-PS, Health and Education, Microsoft India, said, 'With AI being the defining technology of our time, it is transforming lives and industry and the jobs of tomorrow will require a different skillset. This will require more collaborations and training and working with AI. That's why it has become more critical than ever for educational institutions to integrate new cloud and AI technologies. The program is an attempt to ramp up the institutional set-up and build capabilities among the educators to educate the workforce of tomorrow.' The program aims to build up the cognitive skills and in-depth understanding of developing intelligent cloud connected solutions for applications across industry. Earlier in April this year, the company announced Microsoft Professional Program In AI as a learning track open to the public. The program was developed to provide job ready skills to programmers who wanted to hone their skills in AI and data science with a series of online courses which featured hands-on labs and expert instructors as well. This program also included developer-focused AI school that provided a bunch of assets to help build AI skills.

(source: analyticsindiamag.com) and the summarized text with *two lines* as output is:

Envisioned as a three-year collaborative program, Intelligent Cloud Hub will support around 100 institutions with AI infrastructure, course content and curriculum, developer support, development tools and give students access to cloud and AI services. The company will provide AI development tools and Azure AI services such as Microsoft Cognitive Services, Bot Services and Azure Machine Learning. According to Manish Prakash, Country General Manager-PS, Health and Education, Microsoft India, said, "With AI being the defining technology of our time, it is transforming lives and industry and the jobs of tomorrow will require a different skillset.

Text Summarization with Hugging Face Transformers: BERT Model

Part 1: Creating a Baseline

This is the first part of a tutorial on setting up a text summarization project. For more context and an overview of this tutorial, please refer to the introduction.

In this part, we will establish a baseline using a very simple 'model, without actually using machine learning (ML). This is a very important step in any ML project, as it allows us to understand how much value ML adds over the time of the project and if it's worth investing in it.

The code for the tutorial can be found in this Github repo.

Data, data, data …

Every ML project starts with data! If possible, we always should use data related to what we want to achieve with a text summarization project. For example, if our goal is to summarize patent applications, we should also use patent applications to train the model. A big caveat for an ML project is that the training data usually needs to be labeled. In the context of text summarization, that means we need to provide the text to be summarized as well as the summary (the 'label'). Only by providing both can the model learn what a 'good' summary looks like.

In this tutorial, we will use a publicly available dataset, but the steps and code remain exactly the same as if we use a custom/private dataset. And again, if you have an objective in mind for your text summarization model and have corresponding data, please use your data instead to get the most out of this.

The data we will use is the arXiv dataset which contains abstracts of arXiv papers as well as their titles. For our purpose, we will use the abstract as the text we want to summarize and the title as the reference summary. All the steps of downloading and preprocessing the data can be found in this notebook. The dataset was developed as part of this paper and is licensed under the Creative Commons CC0 1.0 Universal Public Domain Dedication.

Note that the data is split into three datasets: training, validation, and test data. If you'd like to use your own data, make sure this is the case too. Just as a quick reminder, this is how we will use the different datasets:

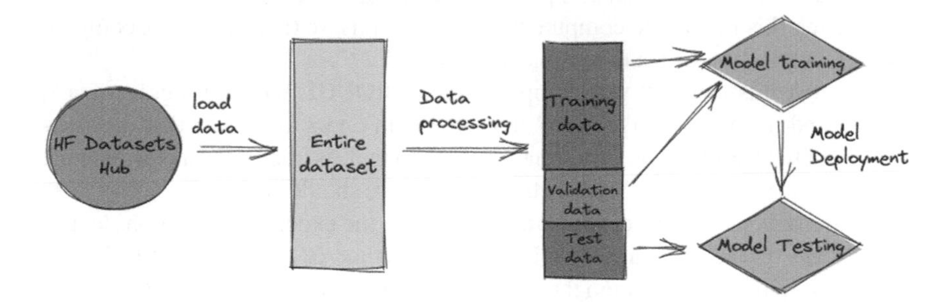

Image by author

Naturally, a common question at this point is: How much data do we need? And, as you can probably already guess, the answer is: It depends. It depends on how specialized the domain is (summarizing patent applications is quite different from summarizing news articles), how accurate the model needs to be useful, how much the training of the model should cost, etc. We will return to this question at a later point when we actually train the model, but the short of it is that we will have to try out different dataset sizes once we are in the experimentation phase of the project.

What makes a good model?

In many ML projects, it is rather straightforward to measure a model's performance. That's because there is usually little ambiguity around whether the model's result is correct. The labels in the dataset are often binary (true/false, yes/no) or categorical. In any case, it's easy in this scenario to compare the model's output to the label and mark it as correct or incorrect.

When generating text, this becomes more challenging. The summaries (the labels) we provide in our dataset are only one way to summarize the text. But there are many possibilities to summarize a given text. So, even if the model doesn't match our label 1:1, the output might still be a valid and useful summary. So how do we compare the model's summary with the one we provide? The metric that is used most often in text summarization to measure the quality of a model is the ROUGE score. To understand the mechanics of this metric, I recommend this blog post. In summary, the ROUGE score measures the overlap of n-grams (contiguous sequence of *n* items) between the model's summary (candidate summary) and the reference summary (the label we provide in our dataset). But, of course, this is not a perfect measure, and to understand its limitations, I quite like this post.

So, how do we calculate the ROUGE score? There are quite a few Python packages out there to compute this metric, and to ensure consistency, we should use the same method throughout our project. Because we will, at a later point in this tutorial, be quite smart and use a training script from the Transformers library instead of writing our own, we can just peek into the source code of the script and copy the code that computes the ROUGE score:

By using this method to compute the score, we ensure that we always compare apples to apples throughout the project.

Note that this function will compute several ROUGE scores: rouge1, rouge2, rougeL, and rougeLsum (The 'sum' in rougeLsum refers to the fact that this metric is computed over a whole summary, while rougeL is computed as the average over individual sentences). So, which ROUGE score should we use for our project? Again, we will have to try different approaches in the experimentation phase. For what it's worth, the original ROUGE paper states that 'ROUGE-2 and ROUGE-L worked well in single document summarization tasks' while 'ROUGE-1 and ROUGE-L perform great in evaluating short summaries.'

Creating the baseline

Next up we want to create the baseline by using a simple, no-ML model. What does that mean? Well, in the field of text summarization, many studies use a very simple approach: They take the first *n* sentences of the text and declare it the candidate summary. They then compare the candidate summary with the reference summary and compute the ROUGE score. This is a simple yet powerful approach which we can implement in a few lines of code (the entire code for this part can be found in this notebook):

Note that we use the test dataset for this evaluation. This makes sense because once we train the model, we will also use the same test dataset for the final evaluation. We also try different numbers for *n*, i.e. we start with only the first sentence as the candidate summary, then the first two sentences, and finally the first three sentences.

And these are the results for our first 'model':

```
[4]: import re

ref_summaries = list(df_test['summary'])

for i in range (3):
    candidate_summaries = list(df_test['text'].apply(lambda x: ' '.join(re.split(r'(?<=[.:;])\s', x)[:i+1])))
    print(f"First {i+1} senctence(s): Scores {calc_rouge_scores(candidate_summaries, ref_summaries)}")

First 1 senctence(s): Scores {'rouge1': 31.3, 'rouge2': 15.5, 'rougeL': 26.4, 'rougeLsum': 26.4}
First 2 senctence(s): Scores {'rouge1': 23.9, 'rouge2': 11.4, 'rougeL': 19.2, 'rougeLsum': 19.2}
First 3 senctence(s): Scores {'rouge1': 19.8, 'rouge2': 9.6, 'rougeL': 15.7, 'rougeLsum': 15.7}
```

Image by author

We can see that the scores are highest with only the first sentence as the candidate summary. This means that taking more than one sentence makes the summary to verbose and leads to a lower score. So that means we will use the scores for the one-sentence summaries as our baseline.

It's important to note that, for such a simple approach, these numbers are actually quite good, especially for the rouge1 score. To put these numbers in context, we can check this page, which shows the scores of a state-of-the-art model for different datasets.

Conclusion and what's next

We have introduced the dataset which we will use throughout the summarization project as well as a metric to evaluate summaries. We then created the following baseline with a simple, no-ML model:

	rouge1	rouge2	rougeL	rougeLsum
baseline (first sentence)	31.3	15.5	26.4	26.4
zero-shot model				
fine-tuned model				

Image by author

In the next part, we will be using a zero-shot model, i.e. a model that has been specifically trained for text summarization on public news articles. However, this model won't be trained at all on our dataset (hence the name 'zero-shot').

I will leave it to you as homework to guess on how this zero-shot model will perform compared to our very simple baseline. On the one hand, it will be a much more sophisticated model (it's actually a neural network), and on the other, it's only used to summarize news articles, so it might struggle with the patterns that are inherent to the arXiv dataset.

Part 2: Zero-shot learning

This is the second part of a tutorial on setting up a text summarization project. For more context and an overview of this tutorial, please refer back to the introduction as well as Part 1 in which we created a baseline for our project.

In this blog post, we will leverage the concept of zero-shot learning (ZSL) which means we will use a model that has been trained to summarize text but hasn't seen any examples of the arXiv dataset. It's a bit like trying to paint a portrait when all you have been doing in your life is landscape painting. You know how to paint, but you might not be too familiar with the intricacies of portrait painting.

The code for the entire tutorial can be found in this Github repo. For today's part, we will use this notebook, in particular.

Why zero-shot learning (ZSL)?

ZSL has become popular over the past years because it allows leveraging state-of-the-art NLP models with no training. And their performance is sometimes quite astonishing: The Big Science Research Workgroup has recently released their T0pp (pronounced 'T Zero Plus Plus') model, which has been trained specifically for researching zero-shot multitask learning. It can often outperform models 6× larger on the BIG-bench benchmark and can outperform the 16× larger GPT-3 on several other NLP benchmarks.

Another benefit of ZSL is that it takes literally two lines of code to use it. By just trying it out, we can create a second baseline, which we can use to quantify the gain in model performance once we fine-tune the model on our dataset.

Setting up a zero-shot learning pipeline

To leverage ZSL models, we can use Hugging Face's Pipeline API. This API enables us to use a text summarization model with just two lines of code while it takes care of the main processing steps in an NLP model:

The text is preprocessed into a format the model can understand.
The preprocessed inputs are passed to the model.
The predictions of the model are postprocessed, so you can make sense of them.

It leverages the summarization models that are already available on the Hugging Face model hub.

So, here's how to use it:

That's it, believe it or not. This code will download a summarization model and create summaries locally on your machine. In case you're wondering which model it uses, you can either look it up in the source code or use this command:

When we run this command, we see that the default model for text summarization is called sshleifer/distilbart-cnn-12-6:

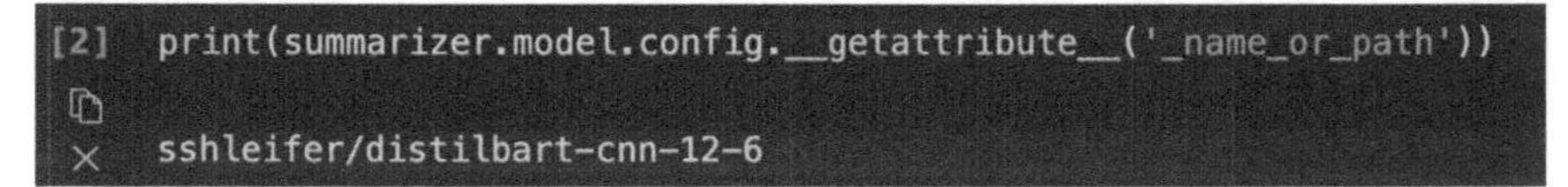

```
[2] print(summarizer.model.config.__getattribute__('_name_or_path'))

sshleifer/distilbart-cnn-12-6
```

Image by author

We can find the model card for this model on the Hugging Face website, where we can also see that the model has been trained on two datasets: the CNN DailyMail dataset and the Extreme Summarization (XSum) dataset. It is worth noting that this model is not familiar with the arXiv dataset and is only used to summarize texts that are similar to the ones it has been trained on (mostly news articles). The numbers 12 and 6 in the model name refer to the number of encoder layers and decoder layers, respectively. Explaining what these are is outside the scope of this tutorial, but you can read more about it in this blog post by Sam Shleifer, who created the model.

We will use the default model going forward, but I encourage you to try out different pretrained models. All the models that are suitable for summarization can be found here. To use a different model, you can specify the model name when calling the Pipeline API:

Sidebar: Extractive vs abstractive summarization

We haven't spoken yet about two possible but different approaches to text summarization: Extractive vs Abstractive. Extractive summarization is the strategy of concatenating extracts taken from a text into a summary, while abstractive summarization involves paraphrasing the corpus using novel sentences. Most of the summarization models are based on models that generate novel text (they are Natural Language Generation models, like, for example, GPT-3). This means that the summarization models will also generate novel text, which makes them abstractive summarization models.

Generating zero-shot summaries

Now that we know how to use it, we want to use it on our test dataset, exactly the same dataset we used in part 1 to create the baseline. We can do that with this loop:

Note that we have the min_length and max_length parameters to control the summary the model generates. In this example we set min_length to 5 because we want the title to be at least five words long. And by eye-balling the reference summaries (i.e. the actual titles for the research papers), it looks like 20 could be a reasonable value for max_length. But again, this is just a first attempt, and once the project is in the experimentation phase, these two parameters can and should be changed to see if the model performance changes.

Sidebar: Beam search, sampling, etc.

If you're already familiar with text generation, you might know there are many more parameters to influence the text a model generates, such as beam search, sampling, and temperature. These parameters give you more control over the text that is being generated, for example make the text more fluent, less repetitive, etc. These techniques are not available in the Pipeline API — you can see in the source code that min_length and max_length are the only parameters that will be considered. Once we train and deploy our own model, however, we will have access to those parameters. More on that in part 4 of this series.

Model evaluation

Once we have the generated the zero-shot summaries, we can use our ROUGE function again to compare the candidate summaries with the reference summaries:

Running this calculation on the summaries that were generated with the ZSL model, we get the following results:

```
[15]: calc_rouge_scores(candidate_summaries, ref_summaries)

[15]: {'rouge1': 30.3, 'rouge2': 14.0, 'rougeL': 26.1, 'rougeLsum': 26.1}
```

Image by author

When we compare those with our baseline from part 1, we see that this ZSL model is actually performing worse than our simple heuristic of just taking the first sentence. Again, this is not unexpected: While this model knows how to summarize news articles, it has never seen an example of summarizing the abstract of an academic research paper.

CONCLUSION

We now have created two baselines, one using a simple heuristic and one with a ZSL model. By comparing the ROUGE scores, we see that the simple heuristic currently outperforms the deep learning model:

	rouge1	rouge2	rougeL	rougeLsum
baseline (first sentence)	31.3	15.5	26.4	26.4
zero-shot model	30.3	14.0	26.1	26.1
fine-tuned model				

Image by author

In the next part we will take this very same deep learning model and try to improve its performance. We will do so by training it on the arXiv dataset (this step is also called fine-tuning): We leverage the fact that it already knows how to summarize text in general. We then show it lots of examples of our arXiv dataset. Deep learning models are exceptionally good at identifying patterns in datasets once they get trained on it, so we do expect the model to get better at this particular task.

Part 3: Training a Summarization Model

In this part we will train the model we used for zero-shot summaries in part 2 (sshleifer/distilbart-cnn-12-6) in our dataset. The idea is to teach the model what summaries for abstracts of research papers look like by showing it many examples. Over time the model should recognize the patterns in this dataset which will allow it to create better summaries.

It is worth noting once more that if you have labeled data, i.e. texts and corresponding summaries, you should use those to train a model. Only by doing so can the model learn the patterns of your specific dataset.

SageMaker training jobs

Because training a deep learning model would take a few weeks on my laptop, we will leverage SageMaker training jobs instead. You can learn all about training jobs in this documentation, but I want to briefly highlight the advantage of using these training jobs, besides the fact that they allow us to use GPU compute instances.

So, let's assume we have a cluster of GPU instances we could use. In that case we would likely want to create a Docker image to run the training so that we can easily replicate the training environment on other machines. We would then install the required packages, and because we want to use several instances, we need to set up distributed training as well. Once the training is done we want to quickly shut down these computers because they are costly.

All these steps are abstracted away from us when using training jobs. In fact, we can train a model in the same way as described above by specifying the training parameters and then just calling one method. SageMaker will take care of the rest, including terminating the GPU instances once the training is completed so to not incur any further costs.

In addition, Hugging Face and AWS have announced a partnership earlier this year that makes it even easier to train Hugging Face models on SageMaker. We can find many examples of how to do so in this Github repo.

Setting up the training job

In fact, we will use one of those examples as a template because it almost does everything we need for our purpose: Training a summarization model on a specific dataset in a distributed manner (i.e. using more than one GPU instance).

One thing, however, we have to account for is that this example uses a dataset directly from HF dataset hub. Because we want to provide our own custom data we need to amend the notebook slightly.

Passing data to training jobs

To account for the fact that we bring our own dataset, we need to leverage channels. You can find more about them in this documentation.

Now, I personally find this term a bit confusing, so in my mind I always think mapping when I hear channels, because it helps me better visualize what happens. Let me try to explain: As we have already learned, the training job spins up a cluster of EC2 instances and copies a Docker image onto it. However, our datasets live in S3 and cannot be accessed by that Docker image. Instead, the training job needs to copy the data from S3 into a predefined path 'locally' into that Docker image. The way it does that is by us telling the training job where the data sits in S3 and where in the docker image the data should be copied into so that the training job can access it. We map the S3 location with the local path.

We set the local path in the hyperparameters section of the training job:

```
[6]: # hyperparameters, which are passed into the training job
     hyperparameters={'per_device_train_batch_size': 4,
                      'per_device_eval_batch_size': 4,
                      'model_name_or_path': 'sshleifer/distilbart-cnn-12-6',
                      'train_file': '/opt/ml/input/data/datasets/train.csv',
                      'validation_file': '/opt/ml/input/data/datasets/val.csv',
```

Image by author

And then we tell the training job where the data resides in S3 when calling the fit() method which will start the training:

```
[8]: huggingface_estimator.fit({'datasets': f's3://{bucket}/summarization/data/'},
```

Image by author

Note that the folder name after/opt/ml/input/data matches the channel name (datasets). This enables the training job to copy the data from S3 to the local path.

Starting the training
Once we have done that, we can start the training job. As mentioned before, this is done by calling the fit() method. The training job will run for about 40 min, and you can follow the progress and see additional information in the console:

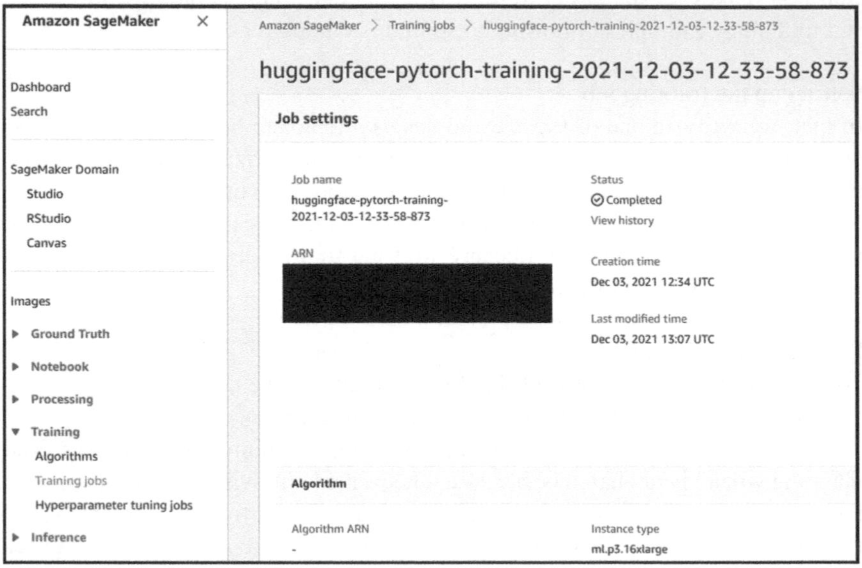

Image by author
The complete code for the model training is in this notebook. Once the training job is finished, it's time to evaluate our newly trained model.

Part 4: Model Evaluation
Evaluating our trained model is very similar to what we have done in part 2 where we evaluated the ZSL model: We will call the model and generate candidate summaries and compare them to the reference summaries by calculating the ROUGE scores. But right now the model sits in S3 in a file called model.tar.gz (to find the exact location you can check the training job in the console). So how do we access the model to generate summaries?

Well, we have two options: We can either deploy the model to a SageMaker endpoint or download it locally similar to what happened in part 2 with the ZSL model. In this tutorial I choose to deploy the model to a SageMaker endpoint because it is more convenient, and by choosing a more powerful instance for the endpoint, we can shorten the inference time significantly. That being said, in the Github repo, you will also find a notebook that shows how to evaluate the model locally.

Deploying a model

It's usually very easy to deploy a trained model on SageMaker, see again this example from Hugging Face. Once the model has been trained, we can just call estimator.deploy(), and SageMaker does the rest for us in the background. Because in our tutorial we switch from one notebook to the next, we have to locate the training job and the associated model first, before we can deploy it:

```
[19]: training_job = client.list_training_jobs()['TrainingJobSummaries'][0]['TrainingJobName']
      model_data = sess.describe_training_job(training_job)['ModelArtifacts']['S3ModelArtifacts']
      model_data

[19]: 's3://sagemaker-us-east-1-905847418383/huggingface-pytorch-training-2021-12-01-09-15-45-087/output/model.tar.gz'
```

Image by author

Once we have retrieved the model location, we can deploy it to a SageMaker endpoint:

Deployment on SageMaker is straightforward because it leverages the SageMaker Hugging Face Inference Toolkit, an open-source library for serving transformers models on Amazon SageMaker. We normally don't even have to provide an inference script, the toolkit takes care of that. In that case, however, the toolkit utilizes the Pipeline API again, and as we have discussed in part 2, the Pipeline API doesn't allow us to use advanced text generation techniques such as beam search and sampling. To avoid this limitation, we provide our custom inference script.

First evaluation

For the first evaluation of our newly trained model, we will use the same parameters as in part 2 with the zero-shot model to generate the candidate summaries. This allows to make an apple-to-apple comparison:

Comparing the summaries generated by the model with the reference summaries:

```
[37]: calc_rouge_scores(candidate_summaries, ref_summaries)

[37]: {'rouge1': 44.5, 'rouge2': 24.5, 'rougeL': 39.4, 'rougeLsum': 39.4}
```

This is encouraging! Our first attempt to train the model, without any hyperparameter tuning, has improved the ROUGE scores significantly:

	rouge1	**rouge2**	**rougeL**	**rougeLsum**
baseline (first sentence)	31.3	15.5	26.4	26.4
zero-shot model	30.3	14.0	26.1	26.1
fine-tuned model	44.5	24.5	39.4	39.4

Image by author

Second evaluation

Now it's finally time to use some more advanced techniques such as beam search and sampling to play around with the model. You can find detailed explanation of what each of these parameters do in this excellent blog post. So let's try it with a semi-random set of values for some of these parameters:

When running our model with these parameters, we get the following scores:

```
[40]: calc_rouge_scores(candidate_summaries_refined, ref_summaries)

[40]: {'rouge1': 43.9, 'rouge2': 23.9, 'rougeL': 38.7, 'rougeLsum': 38.7}
```

Image by author

So that didn't work out quite as we hoped, and the ROUGE scores have actually gone down slightly. However, don't let this discourage you from trying out different values for these parameters. In fact, this is the point where we finish with the setup phase and transition into the experimentation phase of the project.

Final Conclusion and next steps

We have concluded the setup for the experimentation phase. We have downloaded and prepared our data, created a first baseline with a simple heuristic, created another baseline using zero-shot learning, and then trained our own model and saw a significant increase in performance. Now it's time to play around with every part we created in order to create even better summaries. A few ideas you might want to try:

Preprocessing the data properly, e.g. removing stopwords, punctuations, etc. Don't underestimate this part – in many data science projects, data preprocessing is one of the (if not THE) most important aspects and data scientists typically spend most of their time with this task.

Trying out different models. In our tutorial we used the standard model for summarization (sshleifer/distilbart-cnn-12-6), but as we know there are many more models out there that can be used for this task. One of those might be better for your use case.

Hyperparameter tuning. When training the model, we used a certain set of hyperparameters (learning rate, number of epochs, etc.). These parameters are not set in stone, quite the opposite. You want to change these parameters to understand how they affect your model performance.

Different parameters for text generation. We already did one round of creating summaries with different parameters to utilize beam search and sampling.

Abstractive Text Summarization with Bidirectional LSTM

```
# This Python 3 environment comes with many helpful analytics libraries installed
# It is defined by the kaggle/python Docker image: https://github.com/kaggle/docker-python
# For example, here's several helpful packages to load
import numpy as np # linear algebra
import pandas as pd # data processing, CSV file I/O (e.g. pd.read_csv)
# Input data files are available in the read-only "../input/" directory
# For example, running this (by clicking run or pressing Shift+Enter) will list all files
under the input directory
import os
for dirname, _, filenames in os.walk('/kaggle/input'):
      for filename in filenames:
          print(os.path.join(dirname, filename))
# You can write up to 20GB to the current directory (/kaggle/working/) that gets preserved
as output when you create a version using "Save & Run All"
# You can also write temporary files to /kaggle/temp/, but they won't be saved outside of
the current session
```

```
#LSTM with Attention
#pip install keras-self-attention
import numpy as np
import pandas as pd
import re
from bs4 import BeautifulSoup
import tensorflow as tf
from matplotlib import pyplot
from tensorflow.keras.preprocessing.text import Tokenizer
from tensorflow.keras.preprocessing.sequence import pad_sequences
from nltk.corpus import stopwords
from sklearn.model_selection import train_test_split
from tensorflow.keras import backend as K
from tensorflow.keras.layers import LSTM, Bidirectional, Concatenate, Input, Embedding,
TimeDistributed, Dense
from tensorflow.keras.models import Model
from tensorflow.keras.callbacks import EarlyStopping
import sys
sys.path.append('../input')
from attention.attention import AttentionLayer
contraction_mapping = {"ain't":"is not","aren't":"are not","can't":"cannot","'cause":"bec
ause","could've": "could have", "couldn't": "could not",
                              "didn't": "did not", "doesn't": "does not", "don't": "do
not", "hadn't": "had not", "hasn't": "has not", "haven't": "have not",
                              "he'd": "he would","he'll": "he will", "he's": "he is",
"how'd": "how did", "how'd'y": "how do you", "how'll": "how will", "how's": "how is",
                              "I'd": "I would", "I'd've": "I would have", "I'll": "I
will", "I'll've": "I will have","I'm": "I am", "I've": "I have", "i'd": "i would",
                              "i'd've": "i would have", "i'll": "i will", "i'll've": "i
will have","i'm": "i am", "i've": "i have", "isn't": "is not", "it'd": "it would",
                              "it'd've": "it would have", "it'll": "it will", "it'll've":
"it will have","it's": "it is", "let's": "let us", "ma'am": "madam",
                              "mayn't": "may not", "might've": "might have","mightn't":
"might not","mightn't've": "might not have", "must've": "must have",
                              "mustn't": "must not", "mustn't've": "must not have",
"needn't": "need not", "needn't've": "need not have","o'clock": "of the clock",
                              "oughtn't": "ought not", "oughtn't've": "ought not have",
"shan't": "shall not", "sha'n't": "shall not", "shan't've": "shall not have",
                              "she'd": "she would", "she'd've": "she would have",
"she'll": "she will", "she'll've": "she will have", "she's": "she is",
                              "should've": "should have", "shouldn't": "should not",
"shouldn't've": "should not have", "so've": "so have","so's": "so as",
                              "this's": "this is","that'd": "that would", "that'd've":
"that would have", "that's": "that is", "there'd": "there would",
                              "there'd've": "there would have", "there's": "there is",
"here's": "here is","they'd": "they would", "they'd've": "they would have",
                               "they'll": "they will", "they'll've": "they will have",
"they're": "they are", "they've": "they have", "to've": "to have",
                              "wasn't": "was not", "we'd": "we would", "we'd've": "we
would have", "we'll": "we will", "we'll've": "we will have", "we're": "we are",
                              "we've": "we have", "weren't": "were not", "what'll": "what
will", "what'll've": "what will have", "what're": "what are",
                              "what's": "what is", "what've": "what have", "when's": "when
is", "when've": "when have", "where'd": "where did", "where's": "where is",
                              "where've": "where have", "who'll": "who will", "who'll've":
"who will have", "who's": "who is", "who've": "who have",
                              "why's": "why is", "why've": "why have", "will've": "will
have", "won't": "will not", "won't've": "will not have",
                              "would've": "would have", "wouldn't": "would not",
"wouldn't've": "would not have", "y'all": "you all",
                              "y'all'd": "you all would","y'all'd've": "you all would
have","y'all're": "you all are","y'all've": "you all have",
                              "you'd": "you would", "you'd've": "you would have",
"you'll": "you will", "you'll've": "you will have",
                              "you're": "you are", "you've": "you have"}
data=pd.read_csv("../input/amazon-fine-food-reviews/Reviews.csv", nrows=100000)
data.drop_duplicates(subset=['Text'],inplace=True)
data.dropna(axis=0,inplace=True)
stop_words = set(stopwords.words('english'))
#Removes non-alphabetic characters:
def text_cleaner(text):
   newString = text.lower()
   newString = BeautifulSoup(newString, "lxml").text
   newString = re.sub(r'\([^)]*\)', '', newString)
```

```
    newString = re.sub('"','', newString)
    newString = ' '.join([contraction_mapping[t] if t in contraction_mapping else t for t
in newString.split(" ")])
    newString = re.sub(r"'s\b","",newString)
    newString = re.sub("[^a-zA-Z]", " ", newString)
    tokens = [w for w in newString.split() if not w in stop_words]
    long_words=[]
    for i in tokens:
        if len(i)>=2:
            long_words.append(i)
    return (" ".join(long_words)).strip()
def summary_cleaner(text):
    newString = re.sub('"','', text)
    newString = ' '.join([contraction_mapping[t] if t in contraction_mapping else t for t
in newString.split(" ")])
    newString = re.sub(r"'s\b","",newString)
    newString = re.sub("[^a-zA-Z]", " ", newString)
    newString = newString.lower()
    tokens=newString.split()
    newString=''
    for i in tokens:
        if len(i)>1:
            newString=newString+i+' '
    return newString
cleaned_text = []
for t in data['Text']:
        cleaned_text.append(text_cleaner(t))
cleaned_summary = []
for t in data['Summary']:
        cleaned_summary.append(summary_cleaner(t))
data['cleaned_text']=cleaned_text
data['cleaned_summary']=cleaned_summary
data['cleaned_summary'].replace('', np.nan, inplace=True)
data.dropna(axis=0,inplace=True)
#Add sostok and eostok at
data['cleaned_summary'] = data['cleaned_summary'].apply(lambda x : 'sostok '+ x + '
eostok')
#Model to summarize the text between 0-10 words for Summary and 0-50 words for Text
max_len_text = 50
max_len_summary = 10
#Select the Summaries and Text between max len defined above
x_tr, x_val, y_tr, y_val = train_test_split(data['cleaned_text'],data['cleaned_summary'
],test_size=0.2,random_state=0,shuffle=True)
#Tokenize the text to get the vocab count
#prepare a tokenizer for reviews on training data
x_tokenizer = Tokenizer()
x_tokenizer.fit_on_texts(list(x_tr))
#convert text sequences into integer sequences (i.e one-hot encodeing all the words)
x_tr   =  x_tokenizer.texts_to_sequences(x_tr)
x_val  =  x_tokenizer.texts_to_sequences(x_val)
#padding zero upto maximum length
x_tr   =  pad_sequences(x_tr, maxlen=max_len_text, padding='post')
x_val  =  pad_sequences(x_val, maxlen=max_len_text, padding='post')
#size of vocabulary ( +1 for padding token)
x_voc  = len(x_tokenizer.word_index) +1
#prepare a tokenizer for reviews on training data
y_tokenizer = Tokenizer()
y_tokenizer.fit_on_texts(list(y_tr))
#convert text sequences into integer sequences (i.e one-hot encodeing all the words)
y_tr   =  y_tokenizer.texts_to_sequences(y_tr)
y_val  =  y_tokenizer.texts_to_sequences(y_val)
y_tr   =  pad_sequences(y_tr, maxlen=max_len_summary, padding='post')
y_val  =  pad_sequences(y_val, maxlen=max_len_summary, padding='post')
y_voc =  len(y_tokenizer.word_index) +1
latent_dim=550  # Encoder

encoder_inputs = Input(shape=(max_text_len,))
#embedding layer
enc_emb = Embedding(x_voc, embedding_dim,trainable=True)(encoder_inputs)
#encoder lstm 1
encoder_lstm1 = LSTM(latent_dim,return_sequences=True,return_state=True,dropout=0.4,re
current_dropout=0.4)
encoder_output1, state_h1, state_c1 = encoder_lstm1(enc_emb)
#encoder lstm 2
```

```
encoder_lstm2 = LSTM(latent_dim,return_sequences=True,return_state=True,dropout=0.4,re
current_dropout=0.4)
encoder_output2, state_h2, state_c2 = encoder_lstm2(encoder_output1)
#encoder lstm 3
encoder_lstm3=LSTM(latent_dim, return_state=True, return_sequences=True,dropout=0.4,recurr
ent_dropout=0.4)
encoder_outputs, state_h, state_c= encoder_lstm3(encoder_output2)
# Set up the decoder, using `encoder_states` as initial state.
decoder_inputs = Input(shape=(None,))
#embedding layer
dec_emb_layer = Embedding(y_voc, embedding_dim,trainable=True)
dec_emb = dec_emb_layer(decoder_inputs)
decoder_lstm = LSTM(latent_dim, return_sequences=True, return_state=True,dropout=0.4,recur
rent_dropout=0.2)
decoder_outputs,decoder_fwd_state, decoder_back_state = decoder_lstm(dec_emb,initial_s
tate=[state_h, state_c])
#attention layer
attn_layer = AttentionLayer(name='attention_layer')
attn_out, attn_states = attn_layer([encoder_outputs, decoder_outputs])
decoder_concat_input = Concatenate(axis=-1, name='concat_layer')([decoder_outputs,
attn_out])
#dense layer
decoder_dense = TimeDistributed(Dense(y_voc, activation='softmax'))
decoder_outputs = decoder_dense(decoder_outputs)
# Define the model
model = Model([encoder_inputs, decoder_inputs], decoder_outputs)
model = Model([encoder_inputs, decoder_inputs], decoder_outputs)
model.summary()
model.compile(optimizer='rmsprop', loss='sparse_categorical_crossentropy',metrics=['accura
cy'])
es = EarlyStopping(monitor='val_loss', mode='min', verbose=1,patience=50)
history=model.fit([x_tr,y_tr[:,:-1]], y_tr.reshape(y_tr.shape[0],y_tr.shape[1], 1)[:,1:]
,epochs=15, callbacks=[es], batch_size=128, validation_data=([x_val,y_val[:,:-1]], y_val
.reshape(y_val.shape[0],y_val.shape[1], 1)[:,1:]))
pyplot.plot(history.history['loss'], label='train')
pyplot.plot(history.history['val_loss'], label='test')
pyplot.legend()
pyplot.show()
pyplot.plot(history.history['accuracy'], label='train')
pyplot.plot(history.history['val_accuracy'], label='test')
pyplot.legend()
pyplot.show()
reverse_source_word_index = x_tokenizer.index_word
reverse_target_word_index = y_tokenizer.index_word
target_word_index = y_tokenizer.word_index
encoder_model = Model(inputs=encoder_inputs,outputs=[encoder_outputs, state_h, state_c])
decoder_state_input_h = Input(shape=(latent_dim,))
decoder_state_input_c = Input(shape=(latent_dim,))
decoder_hidden_state_input = Input(shape=(max_len_text,latent_dim))
dec_emb2= dec_emb_layer(decoder_inputs)
decoder_outputs2, state_h2, state_c2 = decoder_lstm(dec_emb2, initial_state=[decoder_
state_input_h, decoder_state_input_c])
attn_out_inf, attn_states_inf = attn_layer([decoder_hidden_state_input, decoder_outputs2])
decoder_inf_concat = Concatenate(axis=-1, name='concat')([decoder_outputs2, attn_out_inf])
decoder_outputs2 = decoder_dense(decoder_inf_concat)
decoder_model = Model(
  [decoder_inputs] + [decoder_hidden_state_input,decoder_state_input_h,
decoder_state_input_c],
  [decoder_outputs2] + [state_h2, state_c2])
decoder_model.summary()
def decode_sequence(input_seq):
      e_out, e_h, e_c = encoder_model.predict(input_seq)
      target_seq = np.zeros((1,1))
  target_seq[0, 0] = target_word_index['sostok']
      stop_condition = False
  decoded_sentence = ''
      while not stop_condition:
          output_tokens, h, c = decoder_model.predict([target_seq] + [e_out, e_h, e_c])
          sampled_token_index = np.argmax(output_tokens[0, -1])
          sampled_token = reverse_target_word_index[sampled_token_index]
          if(sampled_token!='eostok'):
        decoded_sentence += ' '+sampled_token
      if (sampled_token == 'eostok' or len(decoded_sentence.split()) >=
(max_len_summary-1)):
```

```
        stop_condition = True
      target_seq = np.zeros((1,1))
      target_seq[0, 0] = sampled_token_index
        e_h, e_c = h, c
  return decoded_sentence
def seq2summary(input_seq):
  newString=''
  for i in input_seq:
      if((i!=0 and i!=target_word_index['sostok']) and i!=target_word_index['eostok']):
        newString=newString+reverse_target_word_index[i]+' '
  return newString
def seq2text(input_seq):
  newString=''
  for i in input_seq:
      if(i!=0):
        newString=newString+reverse_source_word_index[i]+' '
  return newString
for i in range(0,50):
  print("Review:",seq2text(x_tr[i]))
  print("Original summary:",seq2summary(y_tr[i]))
  print("Predicted summary:",decode_sequence(x_tr[i].reshape(1,max_len_text)))
  print("\n")
```

WEB LINKS

1. https://www.geeksforgeeks.org/python-text-summarizer/
2. https://thecleverprogrammer.com/2020/12/31/text-summarization-with-python/
3. https://data-flair.training/blogs/machine-learning-text-summarization/
4. https://www.kaggle.com/code/imkrkannan/text-summarization-with-nltk-in-python/notebook
5. https://www.activestate.com/blog/how-to-do-text-summarization-with-python/
6. https://www.analyticsvidhya.com/blog/2020/12/tired-of-reading-long-articles-text-summarization-will-make-your-task-easier/
7. https://www.mygreatlearning.com/blog/text-summarization-in-python/#Steps%20for%20Implementation
8. https://towardsdatascience.com/understand-text-summarization-and-create-your-own-summarizer-in-python-b26a9f09fc70
9. https://github.com/marshmellow77/text-summarisation-project.git
10. https://towardsdatascience.com/setting-up-a-text-summarisation-project-daae41a1aaa3

https://github.com/kaggle/docker-python

Appendix B: Solutions to Selected Exercises

CHAPTER 1

Ex 2:

	Doc1	Doc2	Doc3
Car	44.55	6.6	39.6
Auto	6.24	68.64	0
Insurance	0	53.46	46.98
Best	21	0	25.5

Ex 3:

A tf = 3/3; idf = log(10000/50) = 5.3; tf-idf = 5.3
B tf = 2/3; idf = log(10000/1300) = 2.0; tf-idf = 1.3
C tf = 1/3; idf = log(10000/250) = 3.7; tf-idf=1.2

Ex 4:

(i) Each occurring term has to be assigned a unique number that corresponds to a dimension of the term space. This can be done in several different ways. For example, one could simply take the ordinal of the term in the alphabet or the order of appearance. Using the latter approach, the document vectors with term frequency are presented as follows:

dimensions
1 2 3

Doc1: (3, 0, 0)
Doc2: (3, 1, 0)
Doc3: (1, 1, 1)
Doc4: (2, 3, 0)
Doc5: (0, 0, 1)
A binary vector (consisting of 1's and 0's) would do as well.

(ii) The scaling with TF*IDF means that the number of occurrences of term in the document (term frequency) is multiplied by the inverse document frequency [1]. The IDF represents the informativity (or the specificity) of the term. If the term occurs in every document of the collection, it has low informativity; it is of little use in determining the similarity between two documents, for instance.

IDF is defined according to the formula: IDF(*t*) = log(*N* / *d*(*t*)), where *N* is the number of documents in the collection and *d*(It) is the number of documents in the collection where the term *t* occurs.

We calculate the inverse document frequencies as follows:

IDF('cat') = 0.321928094887362
IDF('dog') = 0.736965594166206
IDF('mouse') = 1.32192809488736

NOTE: Here we employ the logarithm of base 2. You can switch the base of the logarithm by a simple calculation: If you have *n*-based logarithm and you want to convert to 2-based, you simply divide by $\log_n(2)$, i.e. $\log_2(x) = \log_n(x) / \log_n(2)$.

Base 10 logarithms are just as good as these although the values are considerably smaller. The neat thing about $\log_2$ is that, for a computer scientist, it tells how many bits (binary digits) are needed to represent a given number in the binary format (rounding up). For instance, $\log_2(8) = 3$, which means that you need three bits (23 = 8). Similarly, $\log_{10}$ tells how many digits one needs: $\log_{10}(7) = 1.87 \rightarrow 2$ digits, $\log_{10}(7200) = 3.86 \rightarrow 4$ digits.

As a result, the TF*IDF [2]vectors are:

Doc1: (0.97, 0.00, 0.00)
Doc2: (0.97, 0.74, 0.00)
Doc3: (0.32, 0.74, 1.32)
Doc4: (0.65, 2.21, 0.00)
Doc5: (0.00, 0.00, 1.32)

If one wanted to calculate the normalized TF*IDF weights for the terms, the weight of each term in each vector is divided by the length of the vector. The vector length is calculated as follows:

$$|a| = \sqrt{x_2 + y_2 + z_2}$$

|Doc1| = sqrt(0.97^2 + 0.00 + 0.00) = 0.97
|Doc2| = sqrt(0.97^2 + 0.74^2 + 0.00) = 1.21
|Doc3| = sqrt(0.32^2 + 0.74^2 + 1.32^2) = 1.54
|Doc4| = sqrt(0.65^2 + 2.21^2 + 0.00) = 2.30
|Doc5| = sqrt(0.00 + 0.00 + 1.32^2) = 1.32

Finally, the normalized vectors are:

Doc1: (1.00, 0.00, 0.00)
Doc2: (0.79, 0.61, 0.00)
Doc3: (0.21, 0.48, 0.85)
Doc4: (0.28, 0.96, 0.00)
Doc5: (0.00, 0.00, 1.00)

Ex 6:

Term frequency is how many times a term appears in a particular document in your corpus. Document frequency is how many of the documents in your corpus a term appears in (and inverse document frequency is the multiplicative inverse of this number). Together, these two quantities can be used as a measure of how important a term is to a particular document.

For example: Suppose you have a corpus in which each document is a news article, and you are looking at a document about the New Horizons' space mission to Pluto. You want to figure out which terms are important representative terms for this document. The best such terms will probably be ones that occur frequently in the document (high term frequency) – but you can't just use term frequency, because some words like 'today' and 'said' might also occur frequently in the document, but they're not interesting because they occur frequently in a lot of documents (high document frequency, i.e. low inverse document frequency). You want terms that also have a low document frequency (high inverse document frequency) – that is, they don't occur in many of the other documents (like 'Pluto').

Ex 7:

Step 1: Design the Vocabulary

Now we can make a list of all of the words in our model vocabulary.

The unique words here (ignoring case and punctuation) are:

'it'
'was'
'the'
'best'
'of'
'times'
'worst'
'age'
'wisdom'
'foolishness'

That is a vocabulary of ten words from a corpus containing 24 words.

Step 2: Create Document Vectors

The next step is to score the words in each document. The objective is to turn each free text document into a vector that we can use as input or output for a machine learning model. Because we know the vocabulary has ten words, we can use a fixed-length document representation of 10, with one position in the vector to score each word.

The simplest scoring method is to mark the presence of words as a boolean value, 0 for absent and 1 for present. Using the arbitrary ordering of words listed above in our vocabulary, we can step through the first document ('It was the best of times') and convert it into a binary vector.

The scoring of the document would look as follows:

'it' = 1
'was' = 1
'the' = 1
'best' = 1
'of' = 1
'times' = 1
'worst' = 0
'age' = 0
'wisdom' = 0
'foolishness' = 0

As a binary vector, this would look as follows:

[1, 1, 1, 1, 1, 1, 0, 0, 0, 0]
[1, 1, 1, 1, 1, 1, 0, 0, 0, 0]

The other three documents would look as follows:

'it was the worst of times' = [1, 1, 1, 0, 1, 1, 1, 0, 0, 0]
'it was the age of wisdom' = [1, 1, 1, 0, 1, 0, 0, 1, 1, 0]
'it was the age of foolishness' = [1, 1, 1, 0, 1, 0, 0, 1, 0, 1]

All ordering of the words is nominally discarded, and we have a consistent way of extracting features from any document in our corpus, ready for use in modeling. New documents that overlap with the vocabulary of known words, but may contain words outside of the vocabulary, can still be encoded, where only the occurrence of known words is scored and unknown words are ignored.

Ex 8:

A: tf = 3/3; idf = log(10000/50) = 5.3; tf-idf = 5.3
B: tf = 2/3; idf = log(10000/1300) = 2.0; tf-idf = 1.3
C: tf = 1/3; idf = log(10000/250) = 3.7; tf-idf = 1.2

Ex 9:

a	dog	and	cat
2	1	1	1

a	frog
1	1

The vocabulary contains all words used
a, dog, and, cat, frog
The vocabulary needs to be sorted
a, and, cat, dog, frog
Document A: 'A dog and a cat.'
Vector: (2,1,1,1,0)

a	and	cat	dog	frog
2	1	1	1	0

Document B: 'A frog.'
Vector: (1,0,0,0,1)

a	and	cat	dog	frog
1	0	0	0	1

Ex 10:
BoW model:
V ≡ {bark, brown, cat, colour, dog, green, loud, mat, pull, sit, teeth, white}, |V| = 12.

d1 = (0, 0, 1, 0, 0, 0, 0, 1, 0, 1, 0, 1)
d2 = (1, 1, 1, 1, 1, 0, 1, 0, 0, 0, 0, 0)
d3 = (0, 0, 0, 1, 0, 1, 0, 1, 0, 0, 0, 0)
d4 = (0, 0, 1, 0, 1, 0, 0, 1, 1, 1, 1, 0)

Vector space model:

D1: The cat sat on the mat. The cat was white. |D1| = 5
D2: The brown coloured dog was barking loudly at the cat.|D2| = 6
D3: The mat was green in colour. |D3| = 3

D4: The dog pulled the mat with his teeth. The cat still sat on the mat. |D4| = 7

V ≡ {bark, brown, cat, colour, dog, green, loud, mat, pull, sit, teeth, white}, |V| = 12.

Normalized tf: d1 = 1/ 5 (0, 0, 2, 0, 0, 0, 0, 1, 0, 1, 0, 1)

d2 = 1/ 6 (1, 1, 1, 1, 1, 0, 1, 0, 0, 0, 0, 0)
d3 = 1/ 3 (0, 0, 0, 1, 0, 1, 0, 1, 0, 0, 0, 0)
d4 = 1/ 7 (0, 0, 1, 0, 1, 0, 0, 2, 1, 1, 1, 0)

CHAPTER 2

Ex 2:

There are 12 different bigrams (denoting here the whitespace with 'X' to better see it): Xc, Xh,

Xt, at, ca, cu, eX, ha, he, tX, th, ut

- The corpus being 19 characters long, there are 18 bigrams in total. Here are the counts Xc, 2;

Xh, 1; Xt, 1; at, 2; ca, 1; cu, 1; eX, 2; ha, 1; he, 2; tX, 2; th, 2; ut, 1

- 272 = 729 bigrams in total

Ex 3:

Last
week
,
the
University
of
Cambridge
shared
...(truncated)...

Ex 4:

['Walter',
'feeling anxious',
'He',
'diagnosed today',
'He probably',
'best person I know']

Ex 5:

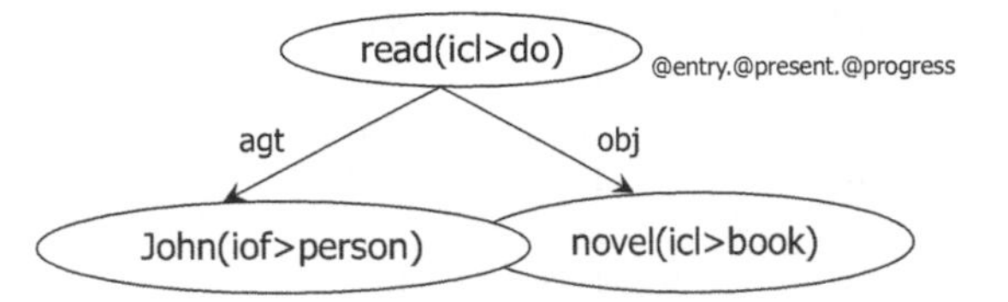


```
Ex 6: {unl}
aoj(director(icl>administrator):07.@entry.@past, he:00)
mod(director(icl>administrator):07.@entry.@past,     academy
                                           (icl>school):0N.@def)
{/unl}
```

Ex 7:

Step 1: Calculate similarity based on the distance function

There are many distance functions, but Euclidean is the most commonly used measure. It is mainly used when data is continuous. Manhattan distance is also very common for continuous variables. The idea to use distance measure is to find the distance (similarity) between new sample and training cases and then find the k-closest customers to the new customer in terms of height and weight.

Euclidean distance between th first observation and the new observation (Monica) is as follows:

$$=SQRT((161 - 158)^2 + (61 - 58)^2)$$

Similarly, we will calculate the distance of all the training cases with new case and calculate the rank in terms of distance. The smallest distance value will be ranked 1 and considered as the nearest neighbor.

Step 2: Find K-nearest neighbors

Let k be 5. Then the algorithm searches for the five customers closest to Monica, i.e. most similar to Monica in terms of attributes, and see what categories those five customers were in. If four of them had 'medium T-shirt sizes' and one had 'large T-shirt size,' then your best guess for Monica is 'medium T-shirt.' Calculation steps are shown in the following table:

Height (in cm)	Weight (in kg)	T-Shirt Size	Distance	Rank of the Nearest Neighbors
158	58	M	4.24	
158	59	M	3.61	
158	63	M	3.61	
160	59	M	2.24	3
160	60	M	1.41	1
163	60	M	2.24	3
163	61	M	2.00	2
160	64	L	3.16	5

163	64	L	3.61
165	61	L	4.00
165	62	L	4.12
165	65	L	5.66
168	62	L	7.07
168	63	L	7.28
168	66	L	8.60
170	63	L	9.22
170	64	L	9.49
170	68	L	11.40

Test data: Height = 161 cm; Weight = 61 kg

Ex 8:

The preprocessing routine is divided as follows:

Corpus tokenization, that is, divide the different texts into individual words.

Stop words removal, which are common words (a, the, not, etc.) that bring close to no contribution to the semantic meaning of a text

Noise removal from the texts, which means basically anything that can't be recognized as an English word, such as words with non-ASCII symbols, words together with numbers, etc.

Stemming, which reduces a word to its root, i.e. [consultant, consulting, consultants] → consult

Feature Extraction with TF-IDF

The preprocessing phase was done so that this stage could yield the best possible results. Here we want to represent how important a word is to a set of documents so that the encoded data is ready to be used by an algorithm. This mapping process of text data into real vectors is known as feature extraction.

TF-IDF, short for term frequency-inverse document frequency, is a numerical statistic that intends to reflect the importance of a word in a corpus, in which a term with a high weight is considered relevant.

This method works by increasing the weight of a word when it appears many times in a document and lowering its weight when it's common in many documents.

This method is divided into two steps:

Term frequency – TF

The term frequency tf(t, d) measures how frequently a term t occurs in a document d, giving a higher weight to more frequent terms.

$$\mathrm{tf}(t,d) = \log\left(1 + \mathrm{freq}(t,d)\right)$$

Inverse document frequency – IDF

The inverse document frequency idf(t, D) measures the importance of a term t in a set of documents D. This statistics lowers the importance of frequent terms with low importance and increases the weight of rare terms that give more meaning to a text.

$$\mathrm{idf}(t,D)=\log\left(\frac{N}{\mathrm{count}(d\in D:t\in d)}\right)$$

The product of tf(t, d) by idf(t, D) yields the tf-idf score for each term.

$$\mathrm{tfidf}(t,d,D)=\mathrm{tf}(t,d)\cdot\mathrm{idf}(t,D)$$

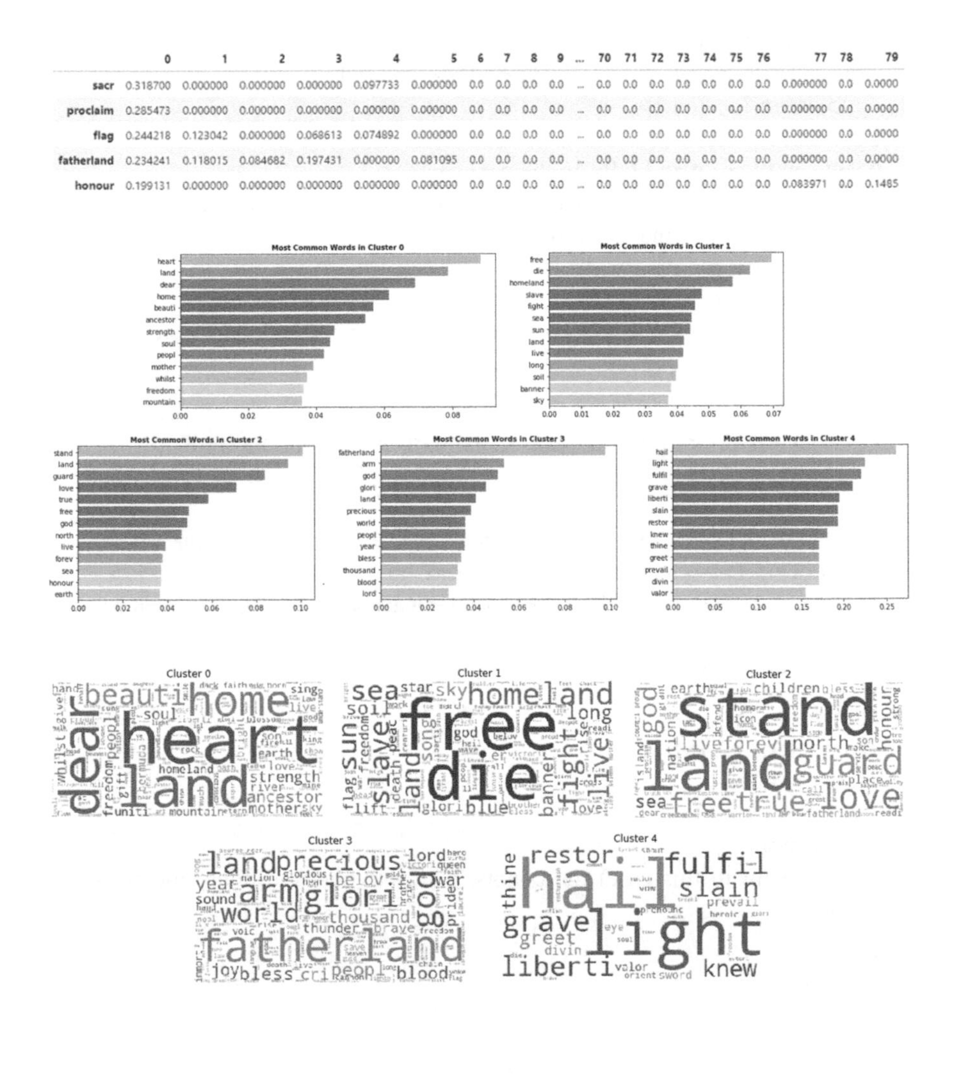

	0	1	2	3	4	5	6	7	8	9	...	70	71	72	73	74	75	76	77	78	79
sacr	0.318700	0.000000	0.000000	0.000000	0.097733	0.000000	0.0	0.0	0.0	0.0	...	0.0	0.0	0.0	0.0	0.0	0.0	0.0	0.000000	0.0	0.0000
proclaim	0.285473	0.000000	0.000000	0.000000	0.000000	0.000000	0.0	0.0	0.0	0.0	...	0.0	0.0	0.0	0.0	0.0	0.0	0.0	0.000000	0.0	0.0000
flag	0.244218	0.123042	0.000000	0.068613	0.074892	0.000000	0.0	0.0	0.0	0.0	...	0.0	0.0	0.0	0.0	0.0	0.0	0.0	0.000000	0.0	0.0000
fatherland	0.234241	0.118015	0.084682	0.197431	0.000000	0.081095	0.0	0.0	0.0	0.0	...	0.0	0.0	0.0	0.0	0.0	0.0	0.0	0.000000	0.0	0.0000
honour	0.199131	0.000000	0.000000	0.000000	0.000000	0.000000	0.0	0.0	0.0	0.0	...	0.0	0.0	0.0	0.0	0.0	0.0	0.0	0.083971	0.0	0.1485

Looking at the clusters, it's clear that the words in each one of them have a theme. In Cluster 0, for example, there are more positive words like 'heart,' 'beauti,' and 'mother, while in Cluster 3, there are more conflict-related words such as 'war,' 'blood,' and 'glori.'

CHAPTER 4

2.

Mapper output:

This is a cat
Cat sits on a roof
<this 1> <is 1> <a <1,1,>> <cat <1,1>> <sits 1> <on 1> <roof 1>
The roof is a tin roof
There is a tin can on the roof
<the <1,1>> <roof <1,1,1>> <is <1,1>> <a <1,1>> <tin <1,1>> <then 1> <can 1> <on 1>
Cat kicks the can
It rolls on the roof and falls on the next roof
<cat 1> <kicks 1> <the <1,1>> <can 1> <it 1> <roll 1> <on <1,1>> <roof <1,1>> <and 1> <falls 1> <next 1>
The cat rolls too
It sits on the can
<the <1,1>> <cat 1> <rolls 1> <too 1> <it 1> <sits 1> <on 1> <cat 1>

Mapper Output:

<this 1> <is 1> <a <1,1,>> <cat <1,1>> <sits 1> <on 1> <roof 1>
<the <1,1>> <roof <1,1,1>> <is <1,1>> <a <1,1>> <tin <1,1>> <then 1> <can 1> <on 1>
<cat 1> <kicks 1> <the <1,1>> <can 1> <it 1> <roll 1> <on <1,1>> <roof <1,1>> <and 1> <falls 1> <next 1>
<the <1,1>> <cat 1> <rolls 1> <too 1> <it 1> <sits 1> <on 1> <cat 1>

Input to the Reducers:

<cat <1,1,1,1>>
<roof <1,1,1,1,1,1>>
<can <1, 1,1>>

Reducer Output:

Reduce (sum in this case) the counts: Can use non-traditional methods for summing

<cat 4>
<can 3>
<roof 6>

3.

1. Map phase: The phase where the individual in-charges are collecting the population of each house in their division is Map Phase.
   Mapper: Involved individual in-charge for calculating population.
   Input splits: The state or the division of the state.
   Key--value pair: Output from each individual Mapper like the key is Rajasthan and value is 2.
2. Reduce phase: The phase where you are aggregating your result
   Reducers: Individuals who are aggregating the actual result. Here in our example, the trained officers. Each reducer produces the output as a key–value pair.
3. Shuffle phase: The phase where the data is copied from mappers to reducers is shuffle phase. It comes in between map and reduces phase. Now the map phase, reduce phase, and shuffler phase are the three main phases of MapReduce.

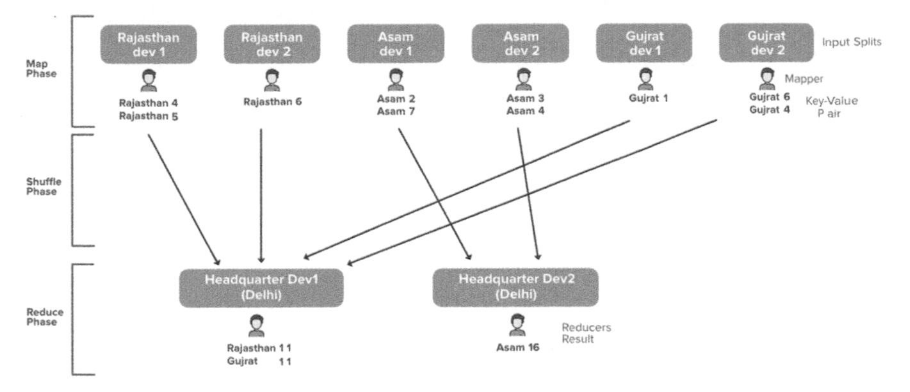

CHAPTER 6

1.

(i) The cases in which the patients actually did not have heart disease and our model also predicted as not having it are called the true negatives. For our matrix, true negatives = 33.

(ii) The cases in which the patients actually have heart disease and our model also predicted as having it are called the true positives. For our matrix, true positives = 43.

(iii) However, there are some cases where the patient actually has no heart disease, but our model has predicted that they do. This kind of error is the type I error, and we call the values false positives. For our matrix, false positives = 8.

(iv) Similarly, there are some cases where the patient actually has heart disease, but our model has predicted that he/she don't. This kind of error is type II error, and we call the values false negatives. For our matrix, false negatives = 7.

Precision = 43/(43 + 8) = 0.843
Recall = 43/(43 + 7) = 0.86
F-score = 2 * (0.843 * 0.86)/(0.843 + 0.86) = 0.85

2. Precision measures the percentage of emails flagged as spam that were correctly classified – that is, the percentage of dots to the right of the threshold line that are green in the figure.
Precision = TPTP + FP = 88 + 2 = 0.8
Recall measures the percentage of actual spam emails that were correctly classified – that is, the percentage of green dots that are to the right of the threshold line in the figure.
Recall = TPTP + FN = 88 + 3 = 0.73
3. For system summary 1:
Recall = 6/6 = 1
Precision = 6/7 = 0.86
F-score = 0.924
For system summary 2:
Recall = 6/6= 1
Precision = 6/11= 0.55
F-score = 0.709
4. System summary bigrams:
the cat,
cat was,
was found,
found under,
under the,
the bed
Reference summary bigrams:
the cat,
cat was,
was under,
under the,
the bed

Based on the bigrams above, the ROUGE-2 recall is as follows:
Recall = 4/5 = 0.8
Precision = 4/6 = 0.67
F-score = 0.729

5.
(i)
Recall = 4/5 = 0.8
Precision = 4/10 = 0.4
F_1 = 2.0*(0.8*0.4) / (0.8 + 0.4) = 0.53333
(ii)
Recall = 0.642
Precision = 0.818
F1 = 0.72

Index

D

E

F

G

H

I

J

K

L

M

U

V

W

X

Z